# AUTHENTIC KNOWING

# AUTHENTIC KNOWING

## *The Convergence of Science and Spiritual Aspiration*

Imants Baruŝs

Purdue University Press
West Lafayette, Indiana

00 99 98 97 96 5 4 3 2 1

♾The paper used in this book meets the minimum requirements of American National Standard for Information Sciences—Permanence of Paper for Printed Library Materials, ANSI Z39.48-1992.

Printed in the United States of America
Design by Anita Noble
Cover photos by J. David Umberger and Vincent P. Walter

**Library of Congress Cataloging-in-Publication Data**

Baruŝs, Imants, 1952–
Authentic knowing : the convergence of science and spiritual aspiration / Imants Baruŝs.
p. cm.
Includes bibliographical references and index.
ISBN 1-55753-084-X (cloth : alk.paper). — ISBN 1-55753-085-8 (paper : alk. paper)
1. Knowledge, Theory of (Religion) 2. Religion and science. 3. Spiritual life. 4. Authenticity (Philosophy) 5. Psychology, Religious. 6. Baruss, Imants, 1952– . I. Title.
BL 51.B255 1996
215—dc20
95-20709
CIP

My keynote, as it were?
It might've been the tendency to
drive toward the root . . .
that from which all comes.

—*Franklin Wolff*

CONTENTS

## FIGURES

# ACKNOWLEDGMENTS

I would like to thank each of the following scholars for the time and care that they took to respond to my requests concerning material falling within their respective areas of expertise: Thomas Langan, University of St. Michael's College, who read and commented on the paragraphs concerning Martin Heidegger's characterization of inauthenticity in chapter 2; Ilona Reid, University of South Australia, who read and commented on chapter 2; Gerry McKeon, University of Western Ontario, who read and commented on the material concerning quantum physics in chapter 3; Edward Wright, St. Francis Xavier University, who read and commented on chapter 3; Brenda Dunne and Roger Nelson, of the Princeton Engineering Anomalies Research Laboratory, Princeton University, who read and commented on the description of their work in chapter 3; Maggy Harsch-Fischbach, who read and commented on the paragraphs concerning instrumental transcommunication in chapter 3; Roger Walsh, College of Medicine, University of California, Irvine, who read and commented on the paragraphs concerning his experiences in chapter 4; Doroethy Leonard, who made available unpublished material, answered numerous questions, and read and commented on the sections concerning Franklin Wolff in chapter 4; Thomas McFarlane, University of Washington, who read and commented on the sections concerning Franklin Wolff's experiences and philosophy in chapter 4; Ron Leonard, University of Nevada, Las Vegas, who read and commented on the sections concerning Franklin Wolff's experiences and philosophy in chapter 4 and participated in writing definitions of the terms used by Wolff for the glossary; Douglas Baker, who read and commented on chapter 5; Charles Tolbert, University of Virginia, who explained to me the precession of the equinoxes described in chapter 5; and Andrew Schneider, who read and commented on chapters 6 and 7. I thank Lynne Jackson, University of Western

Ontario; David Stinson, a goatherd in the Highlands of Nova Scotia; and students in my humanistic psychology class during the 1994–95 academic year, for reading and providing feedback on the entire manuscript. I thank Deborah Stuart for her perceptive criticisms of the manuscript, her research assistance during the final stages of rewriting, and her undeviating support.

I am grateful to King's College, affiliated with the University of Western Ontario, for a paid sabbatical leave from teaching duties as well as for research funds that made it possible to complete this project. I appreciate the opportunity and collegial environment offered by the members of faculty and Ronald Johnson, chairperson of the Department of Psychology, to write this book as a visiting professor at St. Francis Xavier University during the 1993-94 academic year.

I appreciate the wonderful effort by everyone at Purdue University Press, particularly the unwavering enthusiasm and patience of the director, David Sanders, and the managing editor, Margaret Hunt. I am also grateful for the comments and suggestions made by three referees and by members of the editorial board of Purdue University Press.

I thank my research assistants, Marcel Lecker, John Miedema, Louis Dillon, Marilyn Horne, Robert Graham, Nancy MacDonald, Stephanie Owen, and Linda Shier; and my secretaries, Lea Nevett, Julie Siverns, and Phyllis Fidler, for the work that they have done. I have also appreciated the support, comments, and assistance of Jaroslav Havelka, Morgan Gardner, Edīte Ozols, Robert Downie, Rita Barušs, Gail Morningstar, Sharon Villanueva, Don Rhynas, John Clouston, Alan Beck, Tammy Golinski and Carmen Sprovieri.

Finally, and most importantly, I want to say that I have cherished the discussions with my students at King's College, University of Western Ontario, with students at the University of Latvia, and with my colleagues throughout the world who have shared with me their experiences and insights into the meaning of life and the ultimate nature of consciousness and reality.

CHAPTER ONE

# Introduction

## *Authentic Knowing*

> Science . . . has compromised itself, and while it was once the harbor of the truth seeker who escaped religion's dictates and dogmas, now the truth seeker must steadfastly stand apart from science and religion alike.
>
> —*Jane Roberts*

There is a popular story about an encounter by blind people with an elephant. They had never met an elephant before and were trying to figure out what it was. One got hold of the trunk and said that the elephant was like a snake. Another felt the side and thought that it was like a wall. A third pulled on the tail and decided that an elephant was like a rope. None, by herself, could understand what kind of creature this was.[1]

So it is with us, I think, as we encounter the fundamental questions concerning our existence. *Why is anything? What is the meaning of life? What is the nature of reality? How can I know?*

*What should I do?* We may think that we have come across a snake, perhaps deadly to handle. Or a wall that resists our understanding. Or possibly a rope that could provide a means of ascent.

The blind people in this story could also be interpreted as avenues of approach to existential questions. What happens when science broaches the ultimate nature of consciousness and reality? What happens to those devoted to a spiritual way of life? This book is about our scientific and spiritual efforts to understand the meaning of life and to know the nature of consciousness and reality.

In the next two chapters, I examine the ground for our knowledge of existential issues. The second chapter is concerned with social influence and what happens as we try to be true to ourselves. Questions about the nature of science form the subject matter of the third chapter. Then we switch from the ground to transcendent states of consciousness in which, it has been claimed, the existential questions have been resolved. The promise of enlightenment is discussed in the fourth chapter, and a theosophical theory[2] of reality in the fifth. I discuss the process of self-transformation from the ground to the transcendent in the sixth chapter. Finally, in the seventh chapter, authentic knowing is characterized as an outgrowth of both science and spiritual aspiration.

I am not offering mastery of knowledge concerning the difficult issues discussed in this book. Rather, I have tried to remain true to the struggle toward deeper understanding, which often involves tension between polar opposites. There is tension between a scientific approach and one that is religious. There is tension between fact and speculation. And between academic acceptability and personal conviction. In some cases, it is possible to synthesize disparate elements. In other cases, I may only be able to characterize some of the elements and insights that constitute this struggle.

Our central concern is about knowing. We have a tendency sometimes to measure our knowledgeability in terms of the accumulation of information. In that case, any efforts at synthesis are doomed to failure. By the year 2000 there will be more than one million scientific journals in print.[3] How much of this information can anyone master? Even if we were to confine ourselves to the discipline of psychology, it has been estimated that a psychologist can potentially skim read only about 3 percent of the literature in psychology published in a given year.[4] Hence, the emphasis in this book is on understanding. Rather than trying to accumulate as much information as possible, we could see how little reading we need to do in order to understand something.

In trying to facilitate the deepening of understanding, I stay close to the ground of human experience and, in particular, my own. For example, I draw on my own experience when discussing phenomenal aspects of cognitive processes. In this sense I am adopting a phenomenological approach to the subject matter of this book.[5] However, I do this within the academically accepted rational framework so that I can appeal to the logical coherence of my arguments. At the same time, I use metaphors—such as the elephant story—to illustrate some of my contentions. Much of what I say is speculative, and I try to reflect this in the way that I use language. This is particularly the case with the theosophical theory of reality developed in the fifth chapter. The emphasis on phenomenology and rational discourse are characteristic of the disciplines of humanistic psychology and philosophy, respectively.

But I also appeal to the more material evidence in the natural and social sciences. Thus, in some cases, there is quantitative empirical support for the statements that are made. For example, in the third chapter, my discussion concerning the nature of matter is based on the results of experiments in physics, and that concerning the limitations of science is based on experimental and correlational studies in psychology. Archival analyses have been used in examining the written and oral documents concerning the experiences of others, such as those of Franklin Wolff recounted in the fourth chapter.

In some cases, the basis for my contentions resembles strategies used for gathering information in the social sciences, without being formally identified as such. Thus, my witnessing of events at consciousness-raising workshops is similar to participant observation. There have also been unsystematic qualitative observations that I have made of others' behaviors, such as channeling—discussed in the sixth chapter—that resemble case studies.

In the preceding three paragraphs, I have indicated how the material in this book is related to traditional academic research strategies. Questions concerning appropriate types of research themselves get raised in the course of the discussion of authentic knowing. The point here is not to appeal to definitive evidence of a particular sort in order to convince the reader of the verity of my ideas but to be explicit about the type of support for a variety of statements in an effort to encourage cooperative exploration.

The inclusion of such a variety of material has resulted in a number of technical passages that may be difficult to follow. These are usually only a few paragraphs in length and can be skipped over without losing the train of thought. There is also a plethora of terms whose meanings may be unclear. To rectify this, a glossary has been included in which some of the main meanings of specialized words and phrases have been indicated.

In tackling the subject matter of this book, there is a danger of being associated with new-age[6] enthusiasts with uncritical fantasies about the nature of reality. As soon as we turn to an examination of existential issues, we run into a wealth of information that has not been, and in many cases cannot be, subjected to the methods of scrutiny traditionally employed in the sciences. It is a daunting situation from which it is easier to turn away and to do something more "respectable" than to proceed and risk embarrassment and possible academic castigation. However, having decided to advance, we have no choice but to dig in and hope that our scholarly integrity can be seen through the clouds of nonsense that seek to envelop us.

On the other hand, the very scrupulosity demanded of an academic enterprise may serve as an irritation to readers for whom their spiritual effort is of paramount importance. For them, the concerns of academics may seem as meaningful as the squealing of children running amok in a toy store. Since such frenetic behavior is simply irrelevant to the practical resolution of existential crises, why should such readers put up with the desiccated residue of the ineffable flow of life? For them I would propose that scholarship and, in particular, science, is also a part of the great adventure.

But why should I try to simultaneously accommodate both the scientist and the spiritual aspirant? Let me answer that by indicating the motivation for writing this book. It has grown out of teaching undergraduate courses in consciousness, humanistic psychology, and the psychology of religion, in which questions about the nature of consciousness, the meaning of life, and religious experiences are discussed. Because these courses are taught in the context of scientific psychology at a Catholic college, I need to simultaneously address the concerns of the scientifically and the spiritually oriented students without violating the sensibilities of either. Hence I am constantly striving to interweave these two approaches. This does not mean that the two fit neatly with each other. They do not. In fact, it is necessary to develop an open but critical attitude to the subject matter. Having done that, it becomes possible to reexamine science and spiritual aspiration and to see the extent to which they can converge on some of the fundamental questions that we face.

So, in front of us we have the elephant and the groping. And we are ready to grasp a wrinkle or two of the creature's hide. Due to the unexpected aspects of the subject matter, the reader should not be surprised if some of her wrinkles get turned inside out.

CHAPTER TWO

# Authenticity

## *What It Means to Be True to Oneself*

When the disciple is ready, the guru disappears.

—*Gregory C. Bogart*

When I teach a full-year course in humanistic psychology, on the first day of classes I ask students to write an answer to one of two questions that I put on the blackboard: "What is the meaning of life?" or "What is the purpose of my existence?" Sometimes, when the assignment becomes apparent, I hear nervous laughter and exclamations of disbelief. *Are you serious? You're joking, right? You don't actually expect us to write about that, do you? What are we supposed to say?* After a while the students settle down and start writing.

Some of my students' answers to these questions concerning the meaning of life will be discussed later, at the beginning of chapter 4. First, I want to address the social context within which the existential questions are asked. We do not generally take these questions seriously. They are, somehow, in poor taste. Cause for embarrassment. We tend to shrug them off and get on with our lives. And when we are compelled to acknowledge them, we are sometimes not too sure what to do.

I want to start this chapter by talking about the way in which we are bound within the fabric of our society, as a consequence of which the significance of the existential questions remains hidden from us. I will do this by following three separate avenues of approach, taken from existentialism,[1] humanistic psychology, and social psychology. Then I want to consider the quandary of extricating ourselves from the expectations of others in order to be true to ourselves. For a partial resolution, we will turn to a psychological theory known as psychosynthesis, which will, however, lead us to the fundamental problem of guidance, faced by many spiritual aspirants.

**Inauthenticity**

Let me start with some ideas that have been presented by the German existential thinker Martin Heidegger. In the following three paragraphs, I have stayed close to the language that he used in order to retain the flavor of his train of reasoning. While this may lead to some questions concerning the specific meanings of terms, I hope that the key ideas are nonetheless evident.

Heidegger maintained that there are three characteristics of the way in which we find ourselves in the world: idle talk, curiosity, and ambiguity.[2] "Idle talk" refers to the "vocal gossip" and "scribbling" that occur when we rake up information from our environment and pass it along to others. When we do this, our effort is not to uncover something by assisting others to come into a closer relationship with the entity that is talked about but is rather to close off what is said in the talk as such. Idle talk is characterized by an "undifferentiated kind of intelligibility" that makes it accessible to everyone and allows for the possibility of understanding everything without having to come to grips with any of it. Because understanding is supposed to have been reached, idle talk suppresses disputation and new inquiry. As a result, a person becomes "uprooted" and floats along unattached in a condition of increasing "groundlessness" that remains hidden from her by the "obviousness and self-assurance of the average ways in which things have been interpreted."

Curiosity refers to our tendency of "*not tarrying* alongside what is closest," of "abandoning" ourselves to the world through continuous "*distraction.*" We seek novelty and changing encounters, not to know entities through observation and "marvelling" but simply to have known them. We find ourselves "everywhere and nowhere," again "constantly uprooting" ourselves. In this uprooted condition, it is idle talk that determines what "one 'must' have read and seen." Thus "curiosity, for which nothing is closed off, and idle talk, for which there is nothing that is not understood, provide themselves . . . with the guarantee of a 'life' which, supposedly, is genuinely 'lively.'"

But now ambiguity is introduced. We do not know whether something that we hear has been understood through a genuine effort or not. Indeed, this state of ambiguity is carefully protected. The effort to break something down in order to come to grips with it lives at a slower rate than idle talk, so that idle talk has "long since . . . gone on" to whatever is the "newest thing," and that which has been "carried through, has come too late if one looks at that which is newest." Furthermore, when something is newly created, idle talk and curiosity give the semblance of understanding and guarantee immediate obsolescence. Ambiguity refers also to the tendency to make "surmises" and to pass the surmising off as that which is "really happening," while carrying an action through to completion is seen as unimportant. However, the integrity of our lives depends upon our actions, not just on surmising. Hence we keep "*going wrong*" as long as our understanding is out there "where the loudest idle talk and the most ingenious curiosity keep 'things moving,' where . . . everything (and at bottom nothing) is happening." This is not a condition that has resulted from deliberate deception and not one that would ever be acknowledged within our social fabric. But it is a condition in which we find ourselves with regard to our everyday existence.

It may be difficult to identify the presence of idle talk, curiosity, and ambiguity in everyday life, so let us begin by looking at a situation that is usually characterized, at least in part, in terms of its distance from everyday life, namely, a cult.[3] A cult can be defined as a religious movement—usually without previous ties to traditional religious systems, usually directed by a charismatic leader—that is rejected by its environment.[4] A participant in a particular cult may believe, for example, that on some specific day the world will end; that Guru, the leader of the group, is the second coming of the Christ; and that the Ashtar command is waiting behind the asteroids to lift all true believers away from a convulsing planet during the apocalypse. How is it that intelligent adults, often well-off professionals, can end up with such alien ideas?

One characteristic of a cult is talking white; that is, a participant in a cult has to say only those things that are consistent with the beliefs of the cult and refrain from saying anything that is not acceptable to its members. The fact that the world will end, that Guru is the second coming of the Christ, and that the Ashtar command is waiting behind the asteroids is not open to serious questioning. Indeed, there is no need for such questioning because everything has an airtight explanation. *Anyone with any intelligence at all can see that these are the latter days and that the poles will shift, causing tremendous earthquakes and tidal waves. Clearly, too, no one else but the Christ could be so self-sacrificing as to put up with the abysmal lack of faith on the part of humanity and the failure of the chosen few to assist in the great task of saving all the poor deluded souls who have not been touched by grace and placed their trust in the Christ. As for the Ashtar command, for goodness' sake, Guru is herself daily in touch with the federation, and they have assured her that the command stands forever ready to serve suffering humanity.* There are penalties for asking the wrong questions. Clearly, the devotee who does not speak white has lost her faith and failed the cause. She has become the target of satanic forces that have perverted her mind. Efforts need to be made to purge her of these evil influences and restore her to her senses. So she talks white in order to not become a black sheep.

It is easy to see idle talk, curiosity, and ambiguity at work here because the contentions of cult members are so contrary to popular opinions. Cult members simply believe what they are told and pass it on; they do not stop long enough to carefully examine the source of their evidence and they have no idea how much of what Guru says is the result of her struggle to understand what is going on and how much of it she intentionally or unintentionally confabulates.

But is the situation in our everyday lives so different? We generally believe that all events are the result of physical cause-and-effect chains, that scientists have privileged access to truth and know what is or is not real, that the world will not end, and that those who think that the Ashtar command is waiting behind the asteroids need to have their heads examined. Indeed, we largely function in the same way as participants in a cult, accepting explanations that are floating around and talking white in order to remain in the good graces of the people with whom we need to live our lives.

To the extent that the same interpersonal dynamics operate in both our society and cults, it is not surprising, then, that intelligent adults can so readily adopt such alien ideas. After all, the process of making sense of the world has not changed, just the environment from which one rakes up one's ideas. Thus, if we were to redefine cults in terms of unquestioned adherence to social norms, we could say that rather than having joined a cult, the intelligent adult has simply switched from one cult to another.

As a second example of this everyday condition, let us consider what may happen when someone reads about idle talk, curiosity, and ambiguity. The response may be something like: *So what? I've read Heidegger. Get on with it. Tell me something I don't already know.* Such a person may have focused on having read Heidegger and not on the phenomena that Heidegger has sought to identify. She may know Heidegger, but she may not have found that to which her attention is being directed. Heidegger's work gets swept up in the faster rate of idle talk, so that further examination is curtailed, curiosity demands that the person move on to something else, perhaps something more contemporary, and in the end, the nature of that which Heidegger was talking about remains unknown.

For Heidegger, the everyday uprooted state, characterized by idle talk, curiosity, and ambiguity, is an inauthentic mode of existence. By anticipating our death and realizing that we each will have to do our own dying, we can arrest the ongoing tranquilization, ground our understanding in our experience, and choose actions that realize our own possibilities. Thus we can come to live our lives authentically.[5]

Instead of pursuing Heidegger's ideas concerning the movement toward authenticity, let us switch to another line of reasoning and adopt a developmental approach. How does an individual find herself in an inauthentic mode of being in the first place? Can the dynamics of enmeshment reveal anything about the process of extrication? Although he did not know Heidegger's work, Carl Rogers, one of the founders of humanistic psychology, describes psychological processes that overlap the phenomena of inauthenticity and authenticity.[6]

### Ways of Valuing

According to Rogers, every person has an operative "valuing process" that acts in such a way that preference is shown for certain objects and objectives over others.[7] In the case of an infant, she presumably chooses experiences that enhance the organism that she is and rejects those that do not. For example, hunger would be rejected, whereas food would be positively valued until the infant was satiated. The source of the valuing process is within the organism, so that, moment by moment, experiences are weighed and actions selected to "actualize the organism."

What happens to this idyllic state? According to Rogers, an infant needs love and will modify her behavior in order to achieve experiences of love. But this necessarily leads to conflicts between the infant's own desires and the conditions under which love becomes available. What feels good is often bad from another's point of view. The infant gradually learns to distrust her own experiences and to adopt concepts provided by others as her own. Because "these concepts are not based on [her] own valuing," the fluid, changing character of the infant valuing process gets replaced with a fixed and rigid one.

Whether or not the details of this process are correct, this discussion helps to reveal the phenomenon of "introjected value patterns by which we live." According to Rogers, commonly held "introjections" include the beneficial nature of obedience—particularly obedience without questioning—and the accumulation of money, appliances, and facts. Having lost the locus of valuing within ourselves, we become subjected to the norms of the community around us. Rogers calls this state "adult valuing." It would correspond approximately to the inauthentic mode of existence described by Heidegger.

While adult valuing may frequently be the final stage for transformations of the valuing process, "some individuals are fortunate" in that they have developed "further in the direction of psychological maturity." To move from the adult to a more "mature" way of valuing, we need to find out how we feel about things. We need to become aware of our experiences and seek to clarify their meanings. *People are supposed to have fun at parties, but how do I feel about them? Do I enjoy parties? No, I don't like parties. I find them tiring. Then why do I go? Why don't I do something I would enjoy instead?* One learns to trust again one's sensations, feelings, and intuitions, which may contain more wisdom than one's rational faculties. In this way the mature way of valuing resembles the infant's way of valuing.

The mature way of valuing also differs from that of the infant. It is "much more complex." There are numerous memories associated with each experience as well as information about possible outcomes of actions. Thus the conceptual dimension of adult valuing is incorporated in mature valuing, but this time in such a manner that experiences are chosen so as to make one "a richer, more complete, more fully developed person."

When this process of maturation is allowed to take place, an individual tends to move away from pretenses and facades, away from doing the things that she feels she "ought to do," away from trying to live out the "expectations of others," and becomes more expressive of her "real feelings," more self-directed, more flexible in setting her goals, more open to experiences, more sensitive and accepting of others, and more interested in intimate, "fully communicative" relationships with them. Based on his own observations, Rogers maintains that mature people tend toward mutual respect and universal values that enhance humanity: "In our transactions with experience we are again the locus or source of valuing, we prefer those experiences which in the long run are enhancing, we utilize all the richness of our cognitive learning and functioning, but at the same time we trust the wisdom of our organism."

**Compliance**

There is a generational gap between Heidegger's notion of inauthenticity and Rogers's adult way of valuing. Another generation later, we find social psychologists studying the dynamics of interpersonal relationships. Heidegger was a philosopher who analyzed and organized observations from his own experience. Rogers was a psychologist who based his ideas not only on his own experience but on observations that he made in the course of counseling others. Social psychologists are advised by their experiences but depend upon systematic observations of real-life events and experimental manipulations in controlled environments for making statements about the nature of social interactions. What is of interest here is largely the research that has been done by social psychologists concerning influence and compliance.

Suppose that you have come to participate in a psychology experiment, and while you are sitting in a room by yourself filling out some forms, smoke starts to pour through a vent into the room. What would you do? Well, being a reasonable person, you would leave the room, find somebody, and let them know that something was wrong. Suppose, however, that two other people whom you do not know are in the room with you, filling out forms, when smoke starts to come through the vent. What would you do? When this experiment was performed, participants in 62 percent of the three-person groups failed to report the smoke within six minutes, by which time the smoke was thick enough to obscure the participants' vision.[8]

This diffusion of responsibility depends, in part, upon pluralistic ignorance. We look to others to find out what we should think. They are busy filling out their forms. Clearly they do not think that this smoke is anything to worry about. *It must just be some kind of vapor or something.* Others are looking at us to see what they should think. When they see us, we are busy filling out our forms. Clearly we do not appear to think that the smoke is anything to worry about, so they do not think that the smoke is anything to worry about either.

Pluralistic ignorance is an example of an organizing feature of social interactions called "the principle of social proof."[9] We look to others to determine what is correct and what it is that we should do. In general, the greater the degree of ambiguity, the more we rely on others to guide our own thoughts and actions. And given the ambiguity concerning existential questions, we would then expect that we would be particularly susceptible to looking to others for answers.

Within social psychology, the study of social cognition is concerned with the processes of noticing, interpreting, remembering, and using information about the social world. This information is organized using schemata and scripts. "Schemas are . . . organized collections of information about some object. (The term *object* can refer to virtually anything—other people, their traits, physical objects, issues, or even ourselves.)"[10] They provide us with frameworks for interpreting information about the world. Once they have been developed, they exert a powerful effect on the way we process information and on our expectations for the future. In this way they also affect our actions. A script is a schema for events that occur over time, "a cognitive structure that specifies a typical sequence of occurrences in a given situation."[11] An example would be the interaction between a person selling an item in a store and the purchaser. We do not expect the salesperson to pay the purchaser, or the purchaser to bring along and play a musical instrument in the store. Although schemata and scripts have been defined in the context of social interactions, they are in fact used more generally for organizing information about the world.

Constellations of schemata and scripts form belief systems, which act as filters through which the world is experienced and that determine the parameters of our interactions with the world. Using these concepts, we have another way of talking about Heidegger's undifferentiated intelligibility and Rogers's adult valuing. Let us pursue these phenomena futher within the context of social psychology.

Much of our cognitive functioning happens automatically. We are not aware of the strategies that we use for managing and acting on the information that we receive. This applies also to decision making, where we are liable to use *"judgmental heuristics,"*[12] mental shortcuts whereby simplified schemata are used for evaluating complex situations. In general, this state of affairs is adaptive—we could never consciously manage the vast amounts of information that we constantly have to process.[13]

The limitations of our cognitive strategies are revealed most obviously in idiosyncratic situations or when they are deliberately used by others to work against us. Less obviously, they restrict self-expression by creating a pervasive background of interpretation and expectations against which we live out our lives. The question becomes, then, to what extent can the detrimental effects of these cognitive strategies be neutralized?

Let us start with a relatively simple situation. If our tendency is to look to others for guidance in ambiguous circumstances, then we can make a point of trying to become more self-reliant. This would be consistent with both an effort toward authenticity and a mature way of valuing. We try to make our own evaluations of situations and make our own decisions for action. If we are in a room with smoke pouring in through a vent, we can deliberately disregard the opinions of others, consult our own experience, and make up our own minds about what is happening. This gives us greater scope for our actions. *I don't care what the others think, and I may end up looking like a fool, but I'm not going to sit here while smoke fills up this room. I am going to do something about it.* We may be able to act more independently.

Let us consider another situation. A popular technique of influence used by compliance professionals is known as low-balling.

First, we are given an incentive to perform some action. Having been induced to carry it out, we find that it has other benefits, of which we were initially unaware. Once we have become dependent on these benefits, the compliance professional throws the low ball and removes the initial incentive. At this point, carrying out the action may no longer be attractive to us, but because, in part, we tend to act in a manner consistent with previous commitments, we may feel compelled to continue with it.

In buying a computer recently, I was reminded of the pervasive use of this compliance technique. I telephoned a dealer and found out and wrote down everything that I wanted to know about a machine in which I was interested. On the basis of this information, I decided that I liked the deal and called back to buy it. As I was going through the list of attributes and peripherals to which we had previously agreed, the conditions of sale would start to change. *The trackball? Did I say that that price included the trackball? I never said that that price included the trackball. Did I say that the customs' brokerage was included in the shipping fees? We never pay for the brokerage.* The dealer had thrown the low ball. The initial incentive, the low price, had been removed. In the meantime, I had decided that I wanted that particular machine. The tendency would be to allow the momentum of my expectations for the satisfaction of my desires to carry through by purchasing the machine under the new conditions rather than to stop and reassess whether or not the price for the changed deal was still a good one. In this particular case, reempowerment of oneself and a little knowledge about compliance techniques can serve to neutralize the effects of judgmental heuristics.

However, let us consider a more difficult situation, again involving low-balling. Michael Pallak, David Cook, and John Sullivan[14] conducted a study in which they looked at the effect of public commitment on the conservation of natural gas. Research assistants were sent to interview homeowners, who had been divided into three groups. In the control group the research assistants simply obtained permission to access the records of natural gas use from the utility company. In the private commitment group, interviewers explained that they represented a group interested in energy conservation, discussed energy conservation strategies, and said that they wanted to see whether the personal efforts of homeowners would make a difference. Interviewers told the participants in the second group that they hoped to publicize the results but assured them that they would not be personally identified. Homeowners in the third group, the public commitment group, were told the same thing as homeowners in the private commitment group except that, rather than being assured of anonymity, they were told that the researchers hoped to list the names of the participants in the study when the results were publicized.

After one month, the results were predictable. There were no differences in energy consumption between the private commitment group and the control group. Simply asking people to try harder did not do anything. On the other hand, the homeowners in the public commitment group used less natural gas than those in the other two conditions.

At this point, the researchers informed the homeowners that the project had been successful and that the study was completed. The participants in the public commitment group were told that it would not be possible to identify them publicly after all. This was the low ball. Then the researchers continued to monitor the consumption of natural gas for another eleven months. They found that the homeowners in the public commitment group continued their energy savings relative to the other two groups.

What happened? The only difference between the public and private commitment groups was the possibility of being publicly identified as a participant in the study. This was enough to make a difference in energy conservation. Lowered use of energy was continued after the initial incentive was removed. While homeowners may have been aware of their energy savings, they were unlikely to have been aware of the specific manipulation that resulted in those savings. They are unlikely to have said to themselves: *Ah, my name in print. People will see what a good woman I am. What? It won't be printed? Oh dear. How clever. I've been low-balled. I guess I just have to resign myself to using less gas.*

This example illustrates just how easily our cognitive strategies could get us into trouble and how difficult it may be to neutralize the effects of compliance techniques. In trying to be authentic, we are unlikely to be able to become aware of all the motivations for our actions and to gain control over all the schemata and scripts that we use. This leaves us open to influence by others.

Knowledge about the ability to influence others raises ethical questions. Suppose that I were an environmentalist. Would it be morally acceptable to low-ball for energy conservation? Or should I first inform a person about the nature of low-balling, gain their consent, and then low-ball them? In this way, I may be able to partially empower them but may lose the effect. More generally, should the widespread manipulation of the public for the benefit of private interests be offset by the manipulation of the public for its own good? In such a case, who would decide what constitutes the public good? Is it ethical to allow continued public exploitation by compliance professionals for their own interests?

Inauthenticity, adult valuing, and compliance are three characterizations of the impact of social influence on our behavior. Unless we are willing to trust the intentions of those who are manipulating us, in all three cases, the implications are the same. An effort must be made to shift the locus of control[15] away from the social sphere back toward the individual. Rather than living out the expectations of others, we need to seek to be true to ourselves.[16]

### Being True to Oneself

But what exactly does it mean to be true to oneself? *All right, I realize that I don't have to comb my hair to the right or the left, but does that mean that I put a blob of grease in it and make it stand on end, or do I shave it all off? I don't have to become an opera star like my parents want, but does that mean that I become an engineer or a poet? I don't have to go to allopathic physicians, but does that mean that I invest time and money on alternative medical therapies? I don't have to behave like a compliant housewife or a macho male, but does that mean I have to be a macho housewife or compliant male?* While it may be desirable that "the life lived [be] chosen *by the person living it*, not by someone else,"[17] how does one know what to choose?

Making choices is not just a matter of choosing differently. My discussion concerning cults underscores the ease with which one set of norms for making choices can be replaced with another. Sometimes the effort to extricate oneself from one's social fabric seems to lead to greater entrenchment in an alternate ideology, with its own values and guidelines for behavior. While one no longer conforms to previous expectations, the tendency toward conformity and unexamined need for certainty have not changed. They have simply shifted from one context to another. Thus simply making different choices, in and of itself, does not indicate that one is true to oneself.

There is another problem. The simple admonishment to be true to oneself presupposes that what it means to be true to oneself is unambiguous. Existentialists and humanistic psychologists assumed that each person was internally homogeneous. Their point was that we were not just objects in the world that could be treated in a reductionistic manner but that we were beings to be respected in our wholeness. However, while it may be laudable to seek to distinguish ourselves from objects, we do not appear to find ourselves as whole beings. Rather, wholeness may need to be achieved as a synthesis of disparate interior elements. The problem is not just one of properly paying attention, so to speak, but of not knowing to which aspect of ourselves we need to pay attention.

Do I really need to make the situation so complicated? Why must I call into question assumptions about the coherence of individuals? To see the limitations of the wholeness assumption, let us go back to Rogers's notion of the valuing process and consider what may happen in the course of the effort to rid oneself of adult valuing. Rather than acting on introjected value patterns, we seek to rely again on our own experience. We move away from an overemphasis on thinking and move toward deeper immersion in our sensations and emotions. We learn to trust, again, the wisdom of our bodies. Rather than being rational, scientific, and dry, we become intuitional, experiential, and moist. We return the fluid of feeling to our arid lives. In fact, feelings become the arbiters of actions. *If it feels good, do it!* It is not hard to see where this is going. Rogers saw the recovery of one's experience as an aspect of the mature way of valuing. But historically some human-potential enthusiasts returned to an infant's way of valuing, which degenerated into hedonism.

An illustration of this has been given in the writings of Barry Stevens, who, according to Rogers's foreword to their *Person to Person: The Problem of Being Human,* "has somehow achieved in her life a wisdom which seems all too rare in these days."[18] Her initial contributions in *Person to Person* were followed by two books: *Don't Push the River* and *Burst Out Laughing.* Among other things, Stevens has commended Fritz Perls, the founder of Gestalt therapy, for smoking in an auditorium in violation of regulations against smoking, described the resolution of a "triangle" with her sister and brother-in-law that had become a "tangle," and, as an example of spontaneity, recounted ingeniously lying to her son's employer.[19] Is the wisdom of mature valuing to be characterized by antisocial and deceitful behavior?

It could be argued that I am missing the point and that there is nothing wrong with the self-serving nature of Stevens's behavior. Certainly, one could be authentic but behave immorally from the point of view of the rest of society.[20] However, these vignettes from Stevens's writing hint at some of the problems that could attend unbridled self-indulgence. Acting only in one's own best interests can encroach on the best interests of others and lead to situations that cannot be resolved without conflict.[21] In addition, self-serving behavior, whereby others are used for the advancement of one's own desires, ironically perpetuates precisely the kind of dehumanization to which humanists objected.[22] Even some who were at the heart of the development of the human-potential movement have come to regard an emphasis on self-gratifying behavior as counterproductive. For example, Michael Murphy, cofounder of Esalen Institute, which has been devoted to the development of human potential since 1962, has said in an interview thirty years later that "our feelings finally are not reliable indicators of truth" and that "if ever there is a conflict between the two, service takes precedence over personal growth."[23] In other words, the wisdom of bodily and emotive inclinations may not always be particularly wise.

In seeking to return to our experience, we may be acting on our loudest urges, those that call for somatic and emotional pleasures. If we simply act on them, then we are returning to a state akin to the infant's way of valuing. In order to enable the mature way of valuing, we may need to sit back rather than rush forward, try to neutralize the noisy desires, and see what else arises in our experiential stream. *What is in here? What other motives do I have? What am I really like?* If we are to be true to ourselves, we need to know something of who we are.

### Psychosynthesis

How far have we got in this chapter? I started by raising the existential questions. However, rather than answering them, I discussed the social context within which we interpret our experiences. Inauthenticity, adult valuing, and compliance all characterize the manner in which we live out the expectations of others. Sometimes, as in the case of cult membership, we switch from the norms provided by society at large to those provided by a specific group. From studies in social psychology, it also became apparent that we may be completely unaware of the manner in which our beliefs and subsequent behaviors can be manipulated. Nevertheless, it could be argued that maturity consists in removing, as much as possible, the locus of control from the social fabric around us and reempowering ourselves. That is to say, we should be true to ourselves.

While it is relatively easy to characterize the state of submission to social expectations, it is hard to know what it means to be true to oneself or how such a state is to be achieved. While acting on bodily and emotive urges may serve to recover one's power, it may also result in regression to an infant's way of valuing. However, such feelings are not the only contents of one's experiential stream. In order to characterize our inner life and to organize the dynamics associated with a multitude of inner prompts, let us use a theoretical perspective known as psychosynthesis.

At the turn of this century, machines that operated using fluid pressure became metaphors for psychological functioning.[24] In these psychodynamic theories, each person is thought of as a large bag of water. If you squeeze the bag in one place, the water has to squish out somewhere else. The most famous of these theories is the psychoanalytic theory of Sigmund Freud. According to him, the squeezing was largely done by libidinal impulses that found their mature form of expression as sexual intercourse.

In 1910, a young Italian psychiatrist, Roberto Assagioli, criticized Freud's model on the grounds that it was only partial. Assagioli maintained that Freud's theory failed to adequately treat the more refined aspects of human nature, such as creativity, will, joy, and wisdom.[25] The psychodynamic theory developed by Assagioli, in which these attributes were integrated, was named "psychosynthesis." In this model Assagioli included the existence of a "higher self" as the source of these more refined elements.[26]

The main constituents of the psyche are given in figure 1, "Assagioli's Egg Diagram."[27]

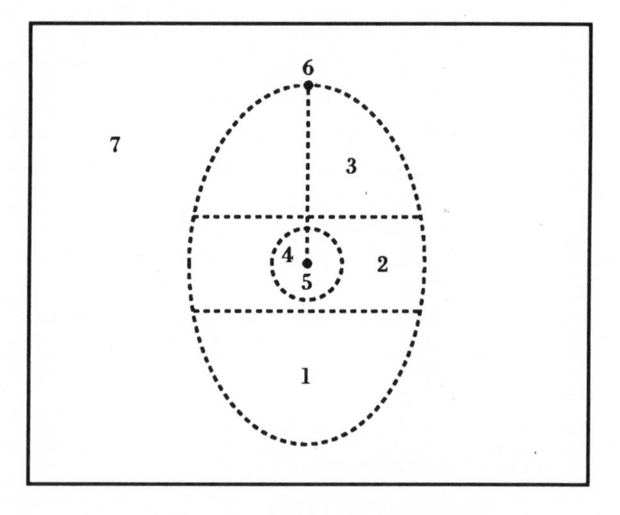


FIG. 1. Assagioli's Egg Diagram

In this diagram, the positions of elements of the psyche are determined by their distance from awareness. Thus, number 1 in figure 1 refers to the subconscious. This is a zone of psychological functioning including cognitive processes controlling the body, primitive drives and urges, emotionally charged complexes that may be manifest as psychopathologies, and some paranormal processes. The subconscious is associated with childhood, is generally the repository of that which was developed in the past, and is largely inaccessible to awareness without some effort. Number 2 refers to the preconscious, the contents of which are readily available to consciousness but are not the focus of attention. For example, the reader may not have been aware of the typeface used in this book until her attention was drawn to it. The superconscious, number 3, is contrasted with the subconscious. The superconscious is the source of altruistic love, ethical imperatives, higher intuition, and inspiration, as well as latent psychic and spiritual qualities. That of which one is directly aware in the present moment lies in the "field of consciousness," number 4. Number 5 is the personal self, the subject for whom experiences occur, "the point of pure self-awareness"[28] at the center of the changing contents of consciousness. Analogously, within the ever changing superconscious lies the unchanging higher self, number 6, of which the personal self is a reflection. Number 7 indicates a collective unconscious, whose elements are shared with all of humanity.[29]

Another important concept taken up by Assagioli was that of subpersonalities.[30] These are unintegrated, circumscribed replications of the egg structure within each person based upon roles played by that individual in her life.[31] There are central subpersonalities, with which the personal self is primarily identified, and peripheral subpersonalities, which are rarely expressed. For example, a woman interested in Buddhist meditation practice who spends her days raising four children may be largely identified with a mother subpersonality and may only at times be identified with her Buddhist monk subpersonality. Furthermore, each subpersonality can become degraded or refined, giving rise to either destructive or benevolent behavior. The more that a subpersonality is acknowledged and refined, the better it can play its unique role within the structure of the personality. The more it is ignored or suppressed, the more likely it is to give rise to destructive or pathological behavior. Subpersonalities are often compared to the players in an orchestra. For the personality to function properly, all of the subpersonalities must cooperate. There is a dynamic quality to this process, so that subpersonalities continue to be formed and integrated during one's lifetime.

Assagioli's theory is an interpretation of experiences that may shed light on our psychological nature. The constructs of the theory, including the notion of the higher self, are ways of organizing phenomena that occur within our stream of consciousness. For Assagioli, the higher self was a real part of the person that could be known directly and clearly.[32] The reader may be uncomfortable with the notion of a real higher self and prefer to think of it as a way of labeling an aspect of specific experiences. In that way, the constructs of the theory can be fruitfully utilized without having to resolve questions about their ontological status.

Psychosynthesis consists not only in a description of psychological dynamics but also of techniques that can be used for revealing and working with the elements of one's psyche.[33] Thus, for example, imagery could be used to evoke subpersonalities. We could visualize a house from which emerge, one by one, a number of characters. These characters could be people, animals, or inanimate objects that symbolize subpersonalities. Some of these subpersonalities may be good at pestering us so that we follow their agendas, while others may be suppressed and unable to contribute to the behavior of the personality. Disidentification is the process of being able to snap out of the illusion that we are a specific subpersonality,[34] to see it for what it is, and to identify with the self.

At the heart of each subpersonality is a quality that is a necessary part of the whole personality. Degraded subpersonalities can be refined. For example, in our imagination, we could walk a subpersonality up the side of a mountain. Such symbolic elevation is thought to lead to refinement of the qualities of a subpersonality. In order to resolve differences among them, we could give subpersonalities opportunities to express themselves and to argue with each other. We could also mediate these arguments by disidentifying from the feuding subpersonalities and adopting as elevated a perspective as possible. In this way, the activity of the subpersonalities may be coordinated, integrated or synthesized, and directed from the self. The term "psychosynthesis" refers to this process of increasingly harmonizing subpersonalities until complete personality integration is achieved.

Humanistic psychologists have placed great value on feelings and emotions. Mentation, particularly rational thought, has not been valued as highly. The exercise of the will has been perceived to be the least meaningful of all. *If it feels good, do it. Don't push the river. Go with the flow.* Assagioli sought to indicate the importance of the will in healthy psychological functioning.[35] He thought that the usual characterization of the will as a brute force that is used to counteract one's desires was a caricature of the will. Rather, the will is a multidimensional, intimate aspect of the self whose function is to decide "what is to be done," to apply "all the necessary means for its realization," and to persist "in the face of all obstacles and difficulties."[36] For example, the will can skillfully direct the activities of the subpersonalities that makes psychosynthesis possible.

So, how far have we got? We have identified, to some degree, inauthenticity, adult valuing, and compliance. Maturity consists in striving to extricate ourselves from living out the expectations of others and being true to ourselves. Because influence occurs at the level of subconscious cognitive strategies that we use in decision making, we cannot always be aware of it. So it may not be possible to guard against all thoughts and actions that we have been manipulated to think and perform. It may not be possible to be entirely self-directed. But what does being true to ourselves mean? We could act on our bodily and emotional urges, but to what extent are they representative of who we are? What about other, less insistent aspects of ourselves? What about our capacity to exercise the will? What about superconscious impulses? What about the higher self? That brings us to another way in which the effort to be true to ourselves can fail.

**Guidance**

We could acknowledge that, indeed, while something may feel good, it may not necessarily be good. That, in fact, rather than listening to our more primitive urges, we should pay attention to spiritual impulses. The model from psychosynthesis provides a schema for the distinction between lower and higher motives. Thus the matter is simple on paper. We have only to listen to the messages coming from the higher self in order to determine what it is that we should do.

What often happens is that the spiritual source becomes objectified. That is to say, who we are spiritually is thought of as being other than that with which we are identified. We may not even think of the spiritual source as a part of ourselves, so that spirit[37] is necessarily other than what we are in essence. We stand, as personalities, over and against that which is spiritual. Then the appropriate script is to try to establish some form of communication between ourselves and the spiritual authority. In traditional religions, gods spoke through intermediaries. Today's aspirant seeking information directly may hear voices, receive intuitive impressions, divine using the I Ching or tarot, or simply interpret the circumstantial events in her life.[38]

Let me give an example of the way in which the last of these might be thought to work. When I first came to teach at King's College, I was given an office with a small window. After spending a decade or so in windowless rooms and the bowels of libraries, a student's dream when she graduates and finally gets a real job is to have an office with a window. So I sat down at my desk and looked out my window. There, directly in my line of vision, was a tower with an air-raid siren. Now, how might I interpret these circumstantial events in my life. If this were a message, what would it be saying? It is obvious, is it not? *Never mind trying to get comfy. There is work to be done.* In the course of the six years that I have been at King's College, the two large trees that frame the air-raid siren have died. I do not know why they have died. But if this were a message, what would it be saying? The previous warning concerning the threat of nuclear war would have been augmented by the threat of environmental degradation.

It is easy to go down a slippery slope and to regard various messages as authoritarian commands. We become disempowered, and the source of the voices or intuitions or divinations or circumstances becomes empowered. We conceptualize ourselves as automata programmed by the spiritual self. We become glorified tubes whose own processes of thinking, evaluation, creativity, judgment, and responsibility are superseded by an anesthetized state in which we simply carry out orders. Rather than living out the expectations of society, we try to live out the expectations of the higher self. We become puppets of god. And we are back to inauthenticity.

Well, what is wrong with this? Nothing, perhaps. However, if we are to regard the tenets of psychosynthesis seriously, then we must consider the possibility that all the elements of the human psyche have a constructive role to play in a person's life. What about our capacity for thinking? What about the development of our own ideas about the world? In fact, the schema of obedience is just one of a number of possible constituents of our belief systems. Perhaps spirituality consists, at least in part, in clarifying what it is exactly that we believe. What about the activity of the will? If the will is a central element of healthy psychological functioning, then lopping it off may remove a necessary ingredient for spiritual growth. Indeed, it may be precisely on the basis of commitment to the results of actions determined by our decisions that we create a process of spiritual deepening.

On the other hand, if there is such a component of our psyche as a higher self, then insisting that judgments be made solely on the basis of information of which we are certain may close off the possibility of superconscious insight. There is a feature of interpersonal relationships that may help to clarify the relationship between the self and a possibly existent, objectified higher self. If one is in a relationship with another human being, one cannot determine one's actions solely on the basis of one's own preferences but must make room for those of the other person as well. Similarly, we could take into account the possibility of impression by the higher self. The dynamics of purported interactions with a higher self will be discussed more fully in chapter 6. For now, let us say that we could adopt an attitude of openness that creates a space for the unknown.

What, then, does it mean to be true to oneself? We have seen that there are two aspects to this. One has to do with disentanglement from social expectations, while the other is concerned with the exercise of our volition. A person may never be able to fully extricate herself from influence, but she can make an effort toward self-directedness. She may want to ground herself in her experience, recognize her inner cacophony, and strive toward personality integration. She can come to trust her own understanding, based on whatever knowledge she has, while open to the unknown. And it may be healthy to exercise her will by doing her own judging, deciding, and acting. Putting all of this together, let us say that authenticity is the effort to act on the basis of one's own understanding.

CHAPTER THREE

# Science

## *What Happens When Science Encounters the Sublime*

Even if it *is* true, I don't believe it.

—*Quoted by Allan Combs and Mark Holland*

We find ourselves with interpretations of reality that we have taken from others, but do the received interpretations work? In particular, a major component of our contemporary culture is science. How well has science dealt with the fundamental questions? What happens when science encounters the sublime? Are we satisfied with the answers? Well, let us start with consciousness. What happens when science encounters consciousness?

## Beliefs about Consciousness and Reality

In reading the academic literature concerning consciousness, one is struck by the confusion surrounding its understanding. One author has said that consciousness does not exist; another, that consciousness is information; and a third, that "pure consciousness" is possible, allowing one to transcend the usual categories for interpreting the world.[1] Why has there been so little agreement about the nature of consciousness? How is one to make sense of these discrepancies? One does not have to look far for an explanation. Different researchers have different beliefs about the nature of reality, and it would not be surprising if these beliefs were to be correlated with their ideas about consciousness. Robert Moore and I examined this hypothesis using the traditional methods of scientific psychology.[2]

In order to investigate the relationship between notions of consciousness and beliefs about reality, a questionnaire was developed. The questionnaire consisted primarily of two sets of items. One set of items was based on statements about consciousness that had appeared in the scientific and philosophical literature. The other set of items was concerned with meaningful beliefs about reality, including statements about the nature of reality, means by which anything can be known, religion, values, extraordinary experiences, intolerance of ambiguity, and existential issues. The questionnaire was distributed to more than 1,491 academics and professionals who could potentially write about consciousness in the academic literature.[3]

Three hundred and thirty-four respondents completed and returned the questionnaires. Of these, 33 percent were associated with the natural and applied sciences, 49 percent with the social and medical sciences, and 18 percent with humanities.[4] More specifically, in terms of disciplinary affiliation, 42 percent of respondents were associated with psychology, 12 percent with physics, 6 percent with philosophy, and smaller numbers with other disciplines. Sixty-seven percent of respondents had obtained a doctorate, and only 4 percent did not have a university degree. The average age was forty-four years, with a standard deviation of twelve years. Twenty-seven percent were women.

Multivariate statistical procedures[5] were used to examine the responses. It was found that three categories of beliefs about reality could be identified. Some respondents tended toward a materialist position, characterized by the belief that reality is entirely physical in nature and that science is the way in which reality is to be known. Other respondents tended toward a conservatively transcendent position. There are two aspects to this position: on the one hand, it is characterized by traditionally religious beliefs; on the other hand, meaning is considered to be important. In either case, both spirituality and meaning attest to the existence of a transcendent reality that cannot be studied scientifically. The third position is an extraordinarily transcendent position, which is comprised of three aspects. First, respondents tending toward this position were more likely to have indicated that they had had extraordinary experiences, such as mystical or out-of-body experiences. Second, this position is characterized by extraordinary beliefs, such as beliefs in reincarnation and extrasensory perception. In fact, physical reality is conceptualized as an extension of mental reality. Third, these respondents emphasized inner experience. Knowledge is to come from introspection, psychological change—possibly brought about by meditation or a spiritual way of life—and modes of understanding superior to rational thought.

As expected, the understanding of consciousness is inherently linked to beliefs about reality. From a materialist point of view, consciousness can be thought of as an emergent property of sufficiently complex neural or information-processing systems. Consciousness is just more information and is always of or about something. Consciousness is not important from this point of view. Rather, emphasis is placed on physiological and computational processes as such. Consciousness is more important from a conservatively transcendent point of view. In fact, the beliefs that "the existence of human consciousness is evidence of a spiritual dimension within each person" and that "consciousness gives meaning to reality" are both aspects of this position. In addition, there is an emphasis on the subjective features of consciousness rather than on emergence or information. From an extraordinarily transcendent point of view, "consciousness is more real than physical reality" and is "the key to personal growth"[6] that is necessary to fully experience and understand human consciousness. In addition to stressing the subjective features of normal modes of consciousness, emphasis is placed on states of consciousness and the existence of a universal consciousness. From the extraordinarily transcendent point of view, consciousness has great importance. The correlation between beliefs about reality and notions of consciousness was judged to be so strong that it was meaningful to talk simply about beliefs about consciousness and reality.[7]

What is striking about these results is the degree to which a materialist-transcendentalist continuum is revealed to exist within the academic community.[8] The materialist pole is characterized by the view that reality is entirely physical in nature, that science is the proper way to understand it, and that consciousness is derived from physical processes. On the other hand, the tendency toward the transcendentalist pole is associated with the beliefs that it is consciousness that is real, that self-transformation can lead to subjective understanding, and that physical reality is derived from consciousness.

So, how does science deal with consciousness? From a materialist point of view, the world is a physical place that functions according to mechanistic laws. No matter how interesting and open to consideration, all phenomena are ultimately to be explained in material terms through a process of reduction. Thus consciousness is believed to be just information, for example. And one is reminded of the "*physical embodiment* principle . . . that information processing occurs in a physical system." For human beings, information depends upon the neurophysiological activity of the brain. In addition, "for purists, the real bottom will of course belong not to neuroscience but to physics." In order to accommodate all the features of consciousness, a materialist may need to tell physicists what properties of matter they will find in the future. This is the sort of thing that can happen when science encounters consciousness.[9]

But what do materialists think that the bottom level of this reduction looks like? What is this matter, of which everything—

including consciousness—is made? What happens when science encounters the material? Do the characteristics of the materialist view, in fact, correctly reflect the nature of reality?

**The Nature of Matter**

Our belief systems are formed of schemata and scripts that we use for organizing the world. As we grow up, in the course of everyday interactions, we are led to develop the impression that the world is made up of substantial objects that interact with one another in a mechanical fashion. We identify sticks, stones, spoons, and dogs as concrete objects whose trajectories can be anticipated when they collide. The concrete objects that we can see are made up of atoms, which are more difficult to observe but nonetheless have the same sorts of properties as those things that can be readily observed. Thus experiences of the world are interpreted using the schema that everything is made up of the analogues of tiny billiard balls that obey immutable, mechanical laws.

But what of these atoms themselves? Of what are they made? How well does this schema apply to subatomic events? It does not; subatomic events cannot be thought of in terms of billiard balls obeying mechanical laws. We need to relinquish our tendency to model subatomic events using the same strategies that we have applied to our everyday experiences.[10] In the domain of the very small, the world does not look or behave the way it does at the people-sized level. In that sense, matter is not material in nature.[11] Yet, despite the fact that this schema has been eroded over the course of the twentieth century, the implications for a materialist interpretation of reality appear to have not been acknowledged by many academics, including physicists, who continue to think of the world as being made up of tiny billiard balls governed by mechanical laws.[12]

What is the nature of subatomic events that makes them incompatible with a billiard-ball description? The prototypical demonstration of the inadequacy of the billiard-ball schema is the two-slit experiment.[13] The setup is such that there is a barrier with two small slits in it that can be opened or closed. On one side of the barrier is an electron gun aimed at the two slits. On the other side of the barrier is a screen that can detect the impact of electrons. Experiments have been conducted in which single electrons, one at a time, have been fired toward the slits and observations made of their impacts on the screen.

Suppose first that one of the two slits is closed and electrons are fired toward the barrier. The detecting screen registers the location of the impact of each electron. Over time, these impacts can be used to define a probability distribution indicating the likelihood of having an electron strike a specific location on the detecting screen. As expected, the highest probability of an impact is across from the slit in line with the electron gun, with the probability dropping off away from the slit. A similar result would be obtained if the other slit were open by itself.

Now, what would happen if single electrons were fired one at a time at the barrier while both slits were open? The result should be straightforward. The impacts of electrons that have gone through one slit would be added to the impacts of those that went through the other slit. The resultant probability distribution should simply be the sum of the two probability distributions determined by closing first one, and then the other, of the two slits.

But that is not what happens. With both slits open, the probability distribution is not just the linear combination of the two single-slit distributions. Rather, in adding up the impacts on the detecting screen and calculating the probability distribution, an interference pattern is seen that is consistent with that associated with wave activity. Even though only one electron at a time was fired at the slits and a discrete impact registered on the detecting screen for each electron, the cumulative distribution of the electrons is inconsistent with the notion that they passed through one or the other of the two slits. The manner in which electrons "travel" from the electron gun to the screen leads one to attribute wavelike qualities to them, which are incompatible with thinking of them as particles.

If we insist on regarding electrons as particles and try to determine through which of the slits individual electrons have gone, then we get a third variant of the two-slit experiment. Suppose that electron detectors are placed at each of the two slits in order to find out through which slit each electron has gone. The electron gun is turned on, both slits with detectors are open, and the detecting screen registers the impact of the electrons. In this case, it can be determined through which slit each electron has gone, and the resultant probability distribution is a simple linear combination of the probability distributions for each slit separately. The act of measuring through which slit an electron has passed has changed the nature of the phenomenon. Now our observations coincide with our intuitions that electrons are particles. But this introduces another, perhaps more serious problem. The dependence of the nature of electrons on the act of measurement is inconsistent with our everyday notion that the behavior of matter is independent of whether or not it is being observed.

From the point of view of everyday ideas about matter, the situation is even worse than this, however, since the decision whether or not to use the electron detectors can be delayed until after an electron has "passed through" the two open slits. If one were to perform such an experiment, one would find that when the electron detectors are used after the electron had "passed through" the slits, it could be determined through which of the two slits the electron had gone, and the resultant probability distribution would be a linear combination of the independent probability distributions for each of the two slits. If the electron detectors were not used, one would find the probability distribution associated with wave interference. It appears as though an eventual decision to detect or not to detect the electron "causes" it to either behave or not behave in a particlelike way at the time that it passes through the two slits. This particular experiment has not been carried out using electrons. However, differently designed experiments using photons have confirmed the delayed-choice phenomenon.[14]

A second set of experiments has also challenged the billiard-ball schema of physical reality. Within quantum theory, which was developed to account for observations of subatomic events, nonlocal effects have been implied. To illustrate this, consider two electrons originating in the same quantum state. This means, among other things, that their properties of spin would be strictly correlated. The spin of one electron would be of equal value to that of the other electron, but in the opposite direction. Suppose now that the two electrons move apart from each other in opposite directions close to the speed of light and do not interact with anything else. When one of these particles is viewed in isolation, it "appears to be undergoing random fluctuations,"[15] so that the spin of the particle cannot be known until it is measured. Quantum theory predicts, however, that once the spin of one electron has been determined through measurement, the spin of the other electron must be of equal value to, and opposite from, that of the measured electron, no matter how far apart the two electrons are at the time of measurement. In other words, subatomic events that were at one time correlated can continue to be correlated in spite of random fluctuations.[16]

Again, the actual experiments to test this have been done with photons rather than electrons, and polarization has been measured rather than spin. The observations of polarization violate those anticipated if the photons had carried the information with them but are consistent with the predictions of quantum theory.[17] Thus strict correlation of some subatomic events is maintained without a materially causal connection between them.

Arthur Zajonc, one of the physicists who has run various delayed-choice experiments, has commented:

> What [these experiments] demand is that we pay closer attention to the theory that we have, and what it means. And to the experiments, because they are surprising—surprising in themselves, as phenomena. . . . [You] try to think through the experiment from beginning to end, as you have been accustomed to in every experimental situation you have ever come across, step by step. And you suddenly realize that your classical ideas are not going to work. But nevertheless, something has happened; this is not just a theory, put together with paper and pencil. Here is an apparatus which requires, for its running, some new idea. And this new idea has a mathematical expression, but it is one which is opaque to common-sense intuition. . . . We are discovering that our intuitions, which were schooled in the physical world of everyday experience, are inadequate to the experiences of the laboratory.[18]

Whatever other implications results of experiments testing quantum theory may have, they demonstrate that our billiard-ball schema about the nature of matter cannot be correct.

Given that matter itself does not conform to our mechanistic intuitions, what is gained by insisting that everything be accounted for in material terms? Why continue to maintain that consciousness be explained in terms of misguided ideas about matter? It may be that, in some ways, matter itself is more mental than material.[19] Physical reality may turn out to be a by-product of consciousness rather than consciousness a by-product of whatever the physical world turns out to be.

Of course, one could try not to descend into the quagmire of matter in the first place and instead give an account of consciousness that does not depend on anything more basic than neural interactions. The problem is that one may not be able to escape from quantum effects, since the size of some of the structures in the brain relevant to neural transmission seem to require that one take into account the nature of subatomic events.[20] If neither the world nor possibly the functioning of the brain can ultimately be explained in material terms, then why should we presume that consciousness is nothing more than the result of material processes?

In our study concerning beliefs about consciousness and reality, we found that the materialist position is characterized by a belief that reality is essentially physical in nature. This belief has been called into question. The materialist position is further characterized by the belief that science is the way in which knowledge is to be acquired. But is science necessarily linked to a materialist interpretation of reality? What is science?

### The Nature of Science

To talk about science as though it were a homogeneous enterprise is misleading. The practice of science varies across disciplines and among individual investigators within disciplines. Furthermore, there is little agreement about what science is or what it should be. In general, scientists themselves learn to do what they do on an apprenticeship basis and have little time or interest to devote to the study of scientific activity as such. This does not stop them, however, from believing that they know what science is, or should be, and being convinced that what they are doing is science. Social scientists, who are generally more "consciously scrupulous" than natural scientists about following the scientific method, appear to be less confident than their colleagues about the legitimacy of what they are doing and tend to advertise the scientific status of their work and disciplines.[21] This heterogeneity needs to be borne in mind when discussing the nature of science.

There are three aspects to science: a worldview, methods of investigation, and an essence. The worldview of science refers to the schemata used for making sense of the world. If the results of quantum physics are taken into account, then there is no agreement about the nature of physical reality. But in the case of some scientists, it would appear that the beliefs described above as a materialist view of the world are also those that define the scientific worldview. Briefly stated, this is the position that the universe is ultimately entirely material in nature and operates in a mechanical way, controlled by fixed laws that are ultimately independent of both human and divine volitional acts.[22]

Perhaps the most salient method of investigation is the scientific method, an idealized set of procedures involving the use of one's sensory modalities for making systematic observations of objective events and drawing inferences from them using one's rational faculties in order to establish general laws.[23] It is portrayed as an unemotional activity, and the investigating scientists are "objective, judicious, disinterested [and] skeptical."[24] The use of this method is thought by some scientists to guarantee that one has established irrevocable truths.[25] According to some, not only is truth established; "the scientific method is 'complete' in the sense that *there do not exist phenomena in nature which, in principle, cannot be explained by application of the scientific method.*"[26]

The essence of science is curiosity about the world. Rather than uncritically accepting authoritative statements, one seeks to acquire knowledge through careful, open-minded investigation from which one can draw conclusions. The process of investigation involves an attitude of mind whereby one sets aside, as much as possible, one's preconceptions about the world and accepts whatever evidence presents itself. The essence of science is thus to turn opinions into questions and then to seek answers to the questions. For example, instead of maintaining that there are no stones in the sky to fall,[27] one asks the question, *Are there stones that can fall from the sky?*

These three aspects of science are causally nested, so that the worldview of science is dependent on the results of the methods of investigation; and the methods of investigation, in turn, are dependent on one's research interests. Thus the opinion that matter at subatomic levels has the same material qualities that appear as attributes of people-sized events had to be given up in the face of experimental evidence. This dynamic was illustrated by Zajonc's comment quoted in the previous section.[28] Similarly, the methods of science may need to be changed depending upon what one is investigating. This requires some further explanation.

There is a Sufi story in which a woman is walking at night when she encounters a man who is searching for something on the ground under a streetlight. "What are you looking for?" the woman asks. "My key," the man replies. She gets down on the ground to help him look for his key. After a while she asks, "Are you certain that you lost it here?" "Oh," says the man, "I know that I didn't lose it here." "Then why are you looking here?" "Because of the light. I can see better."[29] This story illustrates the situation of research being carried out using traditional scientific methods that are inappropriate for the phenomena under investigation.

The limitations of traditional scientific research methods become apparent when one tries to study consciousness. In the study by Moore and myself, we found that those tending toward transcendental positions considered subjective aspects of consciousness to be important. However, in the practice of science, phenomena that are under investigation must be publicly accessible, results must be replicable, and observers must be interchangeable. This means that if one researcher sees that the stuff in the test tube has turned pink, then other researchers had better be able to see that the stuff turns pink under similar conditions. Otherwise one has not established anything. But what happens with subjective experiences that are only privately available?[30] They are accessible, as such, only to the person to whom they occur.[31] Everyone else must infer them from that person's physiological states, behavior, and verbal accounts. The problem is that interchangeability of observers is not possible for events that are subjectively experienced. Hence they fall outside the range of observation through sensory modalities. To the extent that consciousness has subjective features, it, too, falls outside the bounds of traditional scientific investigation.

Psychologists who find themselves within the context of scientific psychology[32] and who may be interested in understanding the subjective aspects of consciousness or the meaning of life are encouraged to find some behavioral aspect of these problems that lies within the scope of accepted methodologies. In this way, they can look where the light is, even though they may know already that they will not find that for which they are looking. In the late 1950s, humanistic psychology grew, in part, out of dissatisfaction with the restrictions that behaviorism placed on the study of psychological processes. Traditional means of measurement were supplemented by accepting the noetic quality of experiences that occurred within one's subjective stream of consciousness.[33] Humanistic psychology, however, while it has had an impact on scientific psychology by bringing attention back to human beings, has generally not been accepted by mainstream psychologists.

If we are serious about coming to know something, then our research methods will have to be adapted to the nature of the phenomenon that we are trying to understand. The purpose of science should take precedence over established methodologies.

By now it is probably clear that the distinction between inauthentic and authentic modes of being, developed in chapter 2, is applicable to science. The inauthentic mode of science has been called "scientism,"[34] while the authentic mode can be called "authentic science." Let us contrast these two modes by briefly considering the inauthentic and authentic manifestations of each of the three aspects of science described above.

The description of the essence of science given above belongs to the authentic mode. In the inauthentic mode, the essence of science is to accumulate a larger and larger pile of facts. The facts, as objective entities, are the valued focus of attention. One serves this pile, and the methods of science are to be strictly enforced as a guarantee that false information does not contaminate the pile. One may not know anything about consciousness but, by golly, one does have pristine facts about the inner ear of the African newt, and this is what is important.

But what meaning do facts, in and of themselves, have? In the nineteenth century, Charles Peirce developed a "Logic of Scientific Discovery,"[35] which included the notion that hypotheses are insights that one seeks to verify within one's experience. The same point was made a century later in the context of consciousness research.[36] The purpose of authentic science is to develop one's understanding, in part, by seeking evidence for one's insights within one's experience.

Similarly, belief in a universal, inflexible scientific method that can guarantee truth belongs to scientism. If one is authentic, one's effort to develop one's understanding by changing opinions into questions may cut so deeply that traditional research methods themselves are called into question and are replaced by others that serve one's purposes better. One may need to draw on the totality of one's experience and not just on that subset that consists of observations made through the process of traditional scientific discovery.

The need to look to one's own experience has been implicitly acknowledged, within the context of research concerning consciousness, by cognitive scientists, who otherwise appear to be committed to traditional scientific values. For example, George Mandler has mentioned that one may test a private theory about one's cognitive functioning against one's direct experience.[37] This appears to be what Daniel Dennett has done in maintaining that upon examination, his computer model of consciousness fits with his experience. He has encouraged others to test his model against their experience in order to assure themselves of its validity.[38] However, the results of an honest investigation cannot be presupposed. What I may find upon examining my experience may not be the same as what Dennett has found upon examining his. Having prescribed the test, he should let me do my own looking and not be surprised if I end up making different observations than he has. For some questions, there may be methods that apparently increase one's understanding without providing universal answers.[39]

The schema that the world is made up of analogues of tiny billiard balls with predictable trajectories largely belongs to an inauthentic worldview of science. From an authentic stance, this schema would be simply a subset of those concepts about the nature of reality that one may use at some point in the course of the development of one's understanding. More generally, to the extent that materialism is adopted without critical examination, it forms part of an inauthentic worldview of science.

The distinction between scientism and authentic science allows us to see that it is scientism rather than authentic science that stands in the way of examining the sublime. That is not to say that the traditional methods of science are inapplicable to a wide range of phenomena. Indeed, while the scientific method may not guarantee truth, the quality of one's knowledge may be improved through its use. The point is not that one must dispense with science; rather, we must relinquish its inauthentic mode in favor of its authentic one.

Thus far, the social dynamics associated with the practice of science have been dealt with implicitly. Let us look at some of them explicitly in order to better understand a scientific approach to the sublime.

### The Social Psychology of Science

William G. Lycan has wondered why it is that "opponents of materialism tend to be simply unmoved by argument *and seem to feel quite justified in being so.*" He has acknowledged also, however, that "materialists remain unembarrassed by the lack of any particularly convincing argument for materialism."[40] There appears to be an assumption that arguments that are presented as being correct should convince others. But they may or may not do so. Similarly, we may be oblivious to empirical evidence. The experiments in physics cited above have been known for some time but have failed to convince many physicists that fundamental ideas about the nature of reality need to be revised. What does it take to convince us?

In the previous chapter, some experiments in social psychology concerning compliance were discussed. More generally, attitudes and attitude change have been studied in social psychology, while the subject of conviction has received less attention.[41] Data from empirical studies have suggested that people discount and counterargue threats to the self, so that threatening parts of messages are processed in a biased manner.[42] Leon Festinger and his colleagues' participant observation of members' behavior in a cult has given some indication of the degree of disconfirmation that must be present before beliefs are given up. Despite repeated failure on the part of extraterrestrials to show up as prophesied on specific dates, there was reluctance on the part of cult members to relinquish their beliefs:

> I've had to go a long way. I've given up just about everything. I've cut every tie: I've burned every bridge. I've turned my back on the world. I can't afford to doubt. I have to believe. And there isn't any other truth.[43]

There has been some evidence to suggest that for beliefs that may not be held with great conviction, clearly contradictory new data can result in the replacement of previous beliefs with ones that are consistent with the data.[44] In other words, there may be a threshold of disconfirmatory evidence that is necessary before beliefs change. This threshold may be dependent on the level of conviction that is associated with those beliefs.[45] While this description is a simplification of some of the dynamics of belief change, it can nonetheless give us some feeling for the role of conviction in the maintenance of beliefs about consciousness and reality.

Thomas S. Kuhn has discussed the way in which anomalies arising in the course of scientific investigation can lead to a crisis whereby the exemplars that one uses for solving similar problems have to be replaced with ones that can adequately account for the anomalous occurrences.[46] Zajonc's comment cited above illustrates the breakdown of accepted exemplars.

However, we are convinced, not only by evidence gathered in the course of scientific investigation but also by events that may occur within the context of our private experience that are outside the reach of traditional scientific methods. Moore and I found that those tending toward the extraordinarily transcendent position were more likely to claim to have had transcendent or mystical experiences, out-of-body experiences, and "experiences which science would have difficulty explaining." In addition, they were more likely to agree with the statement "My ideas about life have changed dramatically in the past."[47] Perhaps these people have had sufficiently clearly disconfirmatory experiences that they have been led to dramatically change their beliefs about life. Whether we like it or not, people's convictions about the nature of reality are influenced by aspects of their experiences other than the processing of arguments and scientifically accumulated facts.

But then, along with the loss of objectivity, universal agreement disappears. Consensus among scientists has been believed to be an attribute of science that elevates it to privileged epistemological status.[48] How, one might ask, is one to deal with the diversity of resultant beliefs if evidence from one's subjective experiential stream is accepted within the domain of science?

This way of characterizing the situation is misleading. One is not admitting or losing anything that has not already been admitted or lost. Admitting evidence from one's subjective experience does not introduce any fundamentally new problems that are not already there to begin with.[49] Objective observations themselves are already experiences within one's experiential stream, and agreement with one's colleagues is a variable rather than a category. In fact, there is better agreement for some subjectively experienced events than there is for putatively objective ones.[50] The only questions, then, are, Where do arbitrary lines get drawn? and Who does the drawing to determine what has fallen over the edge and ceases to be called "science"? However, for the philosopher of mind who finds that his computational model fits his own introspective observations or the psychiatrist who cannot find his own physical body during an out-of-body experience,[51] what difference does it make whether his knowledge is called "scientific"? Both have some degree of conviction regarding the veracity of their understanding of reality based on their own subjective experiences.

Rather than being characterized by consensus, the scientific community is organized in a "lumpy" way with regard to ideas on which there is agreement, illustrated in figure 2.

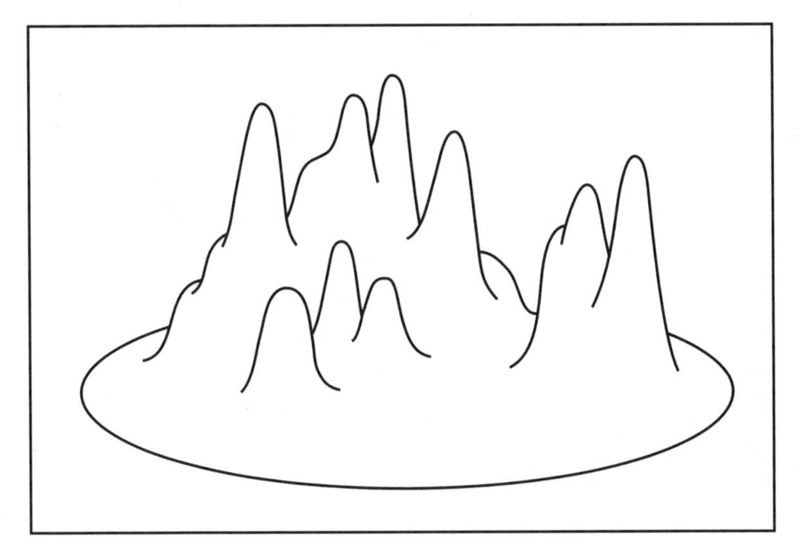

FIG. 2. Organization of the Scientific Community

There are some things on which virtually all scientists would agree, for example, that the sun rises in the east and sets in the west. These would be part of the universal consensus represented by the base of the figure. Movement away from the base indicates decreasing consensus. Thus there are smaller communities of scientists who agree with each other. While they may share some beliefs with other scientists, there are other beliefs that are not shared. For example, there are physicists who believe that seventeen KeV neutrinos exist and others who do not.[52] These two communities of physicists would be represented in the figure by two separate hills, perhaps growing out of a single hill representing the common beliefs of physicists. Ultimately, each investigator is probably on a peak by herself, since no two people have quite the same experiences or interpret similar experiences the same way. In other words, consensus is not a binary quality that is either present or absent for all scientific "facts." Rather, there are subgroups of the scientific community, within which there is agreement with regard to some contentions and not others.[53]

Ironically, consensus is most likely to occur when scientists function inauthentically, uncritically accepting the ideas that they have been taught. Authenticity is characterized precisely by the movement away from compliance by becoming aware of what it is that one actually understands. One could make a distinction between inauthentic and authentic consensus. In the first case, consensus exists due to uncritical acceptance of beliefs. In the second case, consensus occurs when two or more investigators draw the same conclusions. In many cases, the same information can be the content of the locus of inauthentic consensus for some and of authentic consensus for others.

Within traditionally conceptualized science, there are not supposed to be any differences among investigators. The stipulation that observers be interchangeable ensures that only that which is common to all people is applicable to the process of making observations. That which is unique to an individual is eliminated by this requirement. It is a coarse tool with which we probe nature.

But of course, observers are not really interchangeable, nor are unique factors irrelevant to scientific exploration. Shown a photograph of a bubble chamber, I would have no idea which of the swirls was a gluon and which a wino. In fact, it takes someone a decade or so of university training to learn how to make an observation within the scope of her discipline and acceptance by her peers in order to be allowed to run an experiment on a cyclotron.

It becomes evident that the lack of safeguards against fraud and error place greater stress upon professional ethics and the personal integrity of an investigator. A scientist may not be able to rely on external correction but may need to become increasingly scrupulous. Consequently, while it may be possible to resolve discrepancies in some cases—such as the possible production of heat from purported fusion reactions at room temperature—it may be impossible to definitively demonstrate the fictional quality of Carlos Castaneda's accounts of fantastical adventures with magical beings because he was the only accessible witness to these events.[54] However, given the degree of deception concerning events of his personal life that can be verified,[55] one may be reluctant to believe Castaneda's purported anthropological observations, whatever educational value his writing may otherwise have.

Differences in beliefs also have practical consequences. Science is a highly political activity. Scientism still largely dominates the scientific community, and it is difficult for those trying to work in frontier areas of science to obtain funding or have their work published in peer-reviewed academic journals. Not only is innovation discouraged; so is open inquiry. The problem is so extensive that a number of organizations have been founded to provide alternative means of expression for scientists whose work is discredited within the established scientific community. One such organization is the Society for Scientific Exploration, which was founded in 1982 explicitly to provide a forum for scientists whose work falls outside the mainstream of scientific activity.[56] The formation of such organizations, however, does not mitigate the problem of research funding, nor does it help students who may be unable to find programs in which they can pursue their educational objectives.

Despite repeated reminders in the academic literature of the assumptions of the Western mind-set and the limitations of scientism, it is not clear to what extent the situation is actually changing. While it is difficult to generalize about numbers of academics tending toward transcendentalism from the study of beliefs about consciousness and reality by Moore and myself,[57] we were surprised to find so many academics and professionals admitting in an anonymous survey that they had such beliefs. It leads one to think that there may be large numbers of scientists who profess materialist beliefs in order to be able to enjoy an academic career. At some point, perhaps, sufficient numbers of scientists will have had experiences that science cannot explain, creating a critical mass of investigators that could change the political landscape of science.[58]

The following allegory illustrates one version of the possible erosion of scientism. It is nighttime, and there is a campfire in the middle of the woods. A man stands facing the fire and talks.[59] His listeners are sitting in a circle surrounding the fire. As the night goes on and the fire dies down, it is harder and harder for him to see the faces of his listeners. He does not notice that one by one they have slipped away into the night. With the first light of dawn, he sees that the fire has gone out and that he is alone.

**Anomalies**

Schemata associated with materialism work for interpreting reality, as long as we confine ourselves to a limited range of phenomena. As soon as we move toward extremes, these schemata break down. We have seen that matter is not made up of tiny billiard balls and that consciousness cannot be fully explored if we insist on the interchangeability of observers. Even without going to extremes, however, there are anomalous phenomena that do not readily fit into the ordinary worldview of science. In this section, three types of such events are considered: anomalous human-machine interactions, near-death experiences, and instrumental transcommunication. There are numerous other apparent anomalies, such as extrasensory perception, normally functioning hydrocephalic children, out-of-body experiences, synchronicity, mind-body interactions, mental healing, and reincarnation.[60] If one hopes to have an accurate conception of reality, anomalous events have to be taken into account.

Anomalous human-machine interactions have been studied at the Princeton Engineering Anomalies Research Laboratory at Princeton University.[61] A question was raised concerning the vulnerability of sensitive electronic engineering instruments to influence by human consciousness.[62] In order to test this possibility, the following protocol was devised. The signals from a "microelectronic noise source" are "transcribed by appropriate circuitry" in order to provide a "random train of positive and negative pulses."[63] There is a one-way connection from this microelectronic random-event generator to a computer so that the pulses can be symbolically displayed and subjected to statistical analyses. Participants in the study, the operators, record in the computer their intention to get either a greater than random number of positive pulses or a greater than random number of negative pulses or to generate baseline data with "no conscious intention."[64] If the equipment is turned on and allowed to run by itself without operator intention, it really does exhibit random behavior, so that the probability of getting a positive pulse is the same as of getting a negative pulse—50 percent.[65] However, when an operator intends to get more positive pulses than negative pulses, then the probability of getting a positive pulse becomes greater than 50 percent.[66] When an operator intends to get negative pulses, the probability increases in that direction although the shift away from the theoretical expectation is not as great as it is for the intention to get positive pulses.[67] When operators generate a baseline, there are no overall statistically significant changes in the probability of positive and negative pulses.[68] In other words, there appears to be a small, but consistent effect of operator intention on the mean values of the distributions of positive and negative pulses generated by a microelectronic random-event generator.

The question arose whether these types of results could be obtained using macroscopic as well as electronic devices. In order to answer this question, an apparatus called a "random mechanical cascade" was constructed. It consists of a six-foot by ten-foot box with a transparent face, mounted on a wall about eight feet across from a couch. In the upper half of the box are 330 nylon pegs arranged in a quincuncial array, while the lower half consists of 19 collecting bins, each of which is equipped with a photoelectric sensor. Nine thousand ¾-inch polystyrene balls are released, one at a time, through an opening in the top center of the box. As they fall, they bounce off the pegs such that the pattern of the number of balls collected in the bins approximates a normal distribution, with the most balls being in the middle bins and the least number of balls being in the outer bins. An operator, sitting on the couch facing the apparatus, intends to get more of the balls to fall to the right, intends to get more of the balls to fall to the left, or generates a baseline.[69] The result is a statistically significant difference between the means of the distributions for the left-going and right-going intentions.[70] It would appear, then, that one can influence some electronic and macroscopic devices simply by intending to have an effect.

While such anomalous human-machine interactions can be demonstrated using traditional methods of scientific investigation, this is not the case for near-death experiences. A near-death experience is one in which a person is usually close to death around the time of the experience, so that there may be no apparent heartbeat or breathing for some period of time.[71] A person may report having been outside of her body, perhaps looking back down at it, going through darkness into a warm and loving light, and meeting deceased relatives or spiritual beings, perhaps in symbolic forms. Eight million adults in the United States have claimed to have had a near-death experience.[72] Obviously, such phenomena are only privately available for the person who has experienced them. An investigator who has not had a near-death experience herself is restricted to the accounts and physical correlates of such experiences.

It has been argued that near-death experiences result from physiological changes caused in the brain by the traumatic situation of one's body near the time of death or from defensive maneuvers on the part of one's psyche when threatened with nonexistence. However, when Melvin Morse compared 26 children who had nearly died of cardiac arrest with 176 seriously ill children who had been subjected to the same frightening hospital environment and had had "the same lack of oxygen to the brain (as documented by blood tests) and same general blood chemistry," almost all of the children who had been clinically dead had had at least one of the elements of a near-death experience, while none of the other children had had any of these symptoms.[73]

Whatever the explanation for near-death experiences, they have strong effects on those who have had them. First, there is a tendency to believe that death is not the end of life. This is reflected on instruments measuring death anxiety, where lower scores are correlated with the depth of the near-death experience. According to Morse, it may be that experiences of light rather than of being out-of-body determine these changes in death anxiety. Second, despite the gratifying nature of near-death experiences and lessened fear of death, those who have had them tend to value life. In particular, they may be more motivated to improve the quality of their relationships with others and to pursue knowledge that "contributes to the wholeness of the person."[74]

There are some less readily understandable correlates of near-death experiences. For example, those who have reported having had near-death experiences as children are also more likely to have reported having had psychic experiences, such as precognitive dreams.[75] Also, one-quarter of those who have had near-death experiences have claimed that they cannot wear watches because the watches will not run while they wear them.[76] In some cases, they have claimed that this is a transient phenomenon, so that if a dysfunctional watch is placed in a drawer or given to someone else, it resumes proper functioning. Morse has suggested that near-death experiences may change the electromagnetic forces in one's body in such a way that watches and electrical appliances are affected.[77]

There are two strictly logical points that need to be kept in mind by those who would maintain that near-death experiences do not pose a challenge to a materialist point of view. The first point is that only one instance of a genuine nonphysical occurrence is necessary to destroy a materialist philosophy. The materialist view entails the proposition that every phenomenon that anyone has ever experienced can be accounted for in material terms. At present, there is no proof for that general statement. Such a proof would have to demonstrate that every possible experiential event has a material basis. However, in order to show that the statement is false, one needs to produce only a single counterexample. Many of those who have had a near-death experience are convinced that their experience is that counterexample.

The second logical point concerns dismissal of the validity of the knowledge gained during near-death experiences. A materialist may argue that someone who has had a near-death experience is convinced that it occurs in another dimension, outside of her physical body, but that such a conviction is misguided—that these purported events are hallucinations produced by the brain of the person having a near-death experience. Those who have had near-death experiences sometimes make it clear that the near-death experience was not like a hallucination; rather, they report that the events of the near-death experience were as real as those of everyday experiences and, sometimes, that they were more real.[78] Subjective comparisons of sensory perceptions, such as those used to establish Weber's law in psychophysics,[79] have been accepted without question in psychology. If we were to give the same credibility to the introspected comparisons of near-death and everyday experiences, then we have to acknowledge that the reality of near-death experiences is comparable to that of everyday reality. The logical point is simply that if one insists that near-death experiences are hallucinations yet acknowledges that they are experienced to be as real as everyday waking experiences, then everyday waking experiences are also hallucinations. In other words, rather than discrediting the veracity of near-death experiences, one inadvertently undermines materialism itself.

The third anomalous phenomenon that I want to discuss, instrumental transcommunication, is even more controversial than the first two. Instrumental transcommunication purportedly involves two-way communication, via television, computer, or other electronic means, between investigators on the earth and those who are dead.[80] I have included it here not because it has been unequivocally established to be what it appears to be but to give an instance of an unusual phenomenon that can be approached with an investigative attitude. We do not know what research will eventually turn out to reveal—this phenomenon may have a mundane explanation or a more interesting one.

Anomalous human-machine interactions have been studied using traditional scientific methods of inquiry, and near-death experiences can only be studied objectively by gathering accounts of these experiences and looking for behavioral and physiological correlates. Instrumental transcommunication is open to public scrutiny, but the implications of this phenomenon are so dramatic that it has remained outside the scope of mainstream scientific investigation.

One of the pioneers in this work in its contemporary form was Konstantin Raudive, a Latvian psychologist, who, from 1965 until his death in 1974, recorded voices on audiotapes against a background of artificial noises. Purportedly these were the voices of beings who resided in another dimension, some of whom had previously been alive as humans on earth.[81] Since that time, a number of researchers, in independent laboratories in Europe and the United States, have claimed to have established two-way communication with such beings. This communication has included the reception of television images accompanied by voices as well as the generation of text on a computer's hard disk.[82] This ability to communicate has purportedly been developed by collaboration with beings in another dimension, sometimes by following instructions for the building and tuning of electronic equipment. Raudive himself has frequently appeared during communication sessions after his death.[83] Specific information that has been received about the nature of this other dimension indicates that it is responsive to the thoughts of the inhabitants and, in some locations, is similar in appearance to the world in which we now live.[84] However, the structure of the other dimension presumably cannot be explained using our physical theories.[85]

These phenomena are less startling if one takes into account the long history of contacts with another dimension that have purportedly been established through human channels.[86] In fact, instrumental transcommunication entails some of the features of human channeling in that its quality has been said to be dependent on "inner readiness, efforts for a positive spiritual attitude and many other factors we still know little about."[87] Also, as with human channeling, one of the ways of trying to establish the validity of instrumental transcommunication has been to try to verify information provided by a discarnate being that could not reasonably have been known through other means. After George Meek's wife, Jeannette, died in 1990, a message was purportedly received on a computer in Luxembourg containing three items of personal information. Meek could confirm two of these items, but he had no knowledge of the third, the receipt in 1987 by a third person of a box of novels that she had not ordered. However, upon calling that person, he had been able to confirm that she had indeed received such a box of novels.[88]

What are we to make of this? At this point, most of us do not have enough information or experiential evidence to fully understand what is happening. What gives rise to these phenomena? What will further investigation reveal? Will our conceptions about the nature of reality be strengthened, or will they need to be revised? More generally, how well do we take into account various types of anomalous events? How willing are we to change our ideas in the face of evidence?

### Alternative Research Methods

Let us now consider the incorporation of alternative epistemological approaches into authentic science. The need for more liberal research methods has been discussed by a number of authors.[89] Some have sought to find ways of incorporating subjective aspects of experiences, while others have advocated phenomenological methods for the study of psychological events whereby phenomena can be better revealed than they are in ordinary experience.[90] Historically, introspection, the process of inner observation, has been the prototypical method used for studying subjective events. While the ability of a person to have any knowledge about her own mental states has been discredited within cognitive science, it has nonetheless been tacitly accepted by some cognitive scientists.[91] More generally, introspection forms the basis for both quantitative and phenomenological methods of investigating subjective states.[92] However, there is no reason to assume that everyone can make observations of their own mental states with the same degree of facility. Introspection may be a difficult skill that requires extensive training.[93]

In seeking to understand some phenomena that may be of interest, not only does one relax the restrictions against private observations; one also changes one's status from an objective observer to a participant. This has been advocated in anthropology, where one may seek to understand a native culture by adopting its point of view and participating in its rituals. It is also unavoidably the case in particle physics, where one's act of measurement disturbs the system that one is measuring. In order to understand some aspects of consciousness, not only is introspection necessary; indeed, investigators may need to learn to induce, where possible and ethical, the phenomena that they are studying. Participation by the investigator is not without precedent in psychology, where Wilhelm Wundt, often credited as the founder of experimental psychology, had investigators and subjects switch roles in experiments.[94]

Some alterations of consciousness, such as those that occur during near-death experiences, cannot ethically be induced. In some cases, those who have been close to someone who has had a near-death experience may themselves be affected by it. As a result of such contact, they may gain some understanding of such states and share some of the convictions of those who have experienced them. Thus they may come to believe that death is not the end of life or that one can live one's life with a greater sense of appreciation for it. In fact, some of those who "merely become *interested* in near-death experiences" show "a diminished fear of death," increased compassion, "decline in materialist values and heightened ecological concern."[95] Although this has not been systematically demonstrated more generally, it would seem that proximity to anomalous events is a factor in the degree to which such events can change one's understanding of reality.

Given that those who have become interested in near-death experiences become affected by them, are there, more generally, ways of sharing insights concerning anomalous phenomena with those who have not experienced them directly? In the absence of universal consensus in science, each investigator appears to be entirely isolated. But the usual format of presenting and publishing discursive research papers is only one means of communication. Some of the modalities used in the arts may be more effective for conveying a sense of what one has experienced or understood.[96] For example, there has been "increasing interest in metaphor as a form of thought with associated epistemological functions."[97] Improbable as it may seem, could a collective domain of the mind exist where insights may be directly accessible?[98] If such were the case, then more intimate sharing of understanding may be possible.

If one believes that truths are either established or not established by the use of the scientific method, then all statements that one makes resulting from the use of that method have the same degree of support. In communicating one's findings, one has only to demonstrate that one has followed the method properly. But then speculations present problems because they fall outside the range of that which has been established as being true. If the mythical nature of the scientific method becomes apparent, then there is nothing to stop anyone from saying anything. However, one needs to indicate the type and degree of evidence for one's contentions. Thus one may cite the results of experimental studies, with their associated levels of statistical significance, or describe one's experiences as a participant in a cult or simply state that one is speculating and not claim any direct evidence for what one is saying. When a bivalent system is replaced with a multivalent one, then the values need to be made explicit.

The reader may nonetheless be aghast at the relativism implied in figure 2 above illustrating the relativity of consensus among investigators. There have been strong statements made against the use of more liberal methodologies. Ernest R. Hilgard has noted that while these methodologies may be of benefit to investigators with a sense of integrity, they would also encourage those with "free-floating uncritical fantasies about mental life."[99] This is a legitimate concern.

However, there are already lots of free-floating uncritical fantasies about all sorts of things, not the least of which are unexamined materialistic beliefs about the nature of reality. The point is not to avoid uncritical fantasies altogether but to learn to manage them so as not to allow them to interfere with the genuine development of one's understanding. In other words, this is the problem of authenticity. Rather than automatically using the schema *if and only if it is scientific is it true* for directing one's actions, one examines contentions, including the type and degree of evidence for them, in terms of their relevance for one's understanding.

There is still the problem of limited cognitive capacity, however. If tolerance of ambiguity replaces consensus and one carefully examines everyone's ideas, then how does one limit the amount of time and effort spent on free-floating uncritical fantasies? But how does one limit the time and effort spent on "legitimate" contentions? Judgments concerning what to examine and not to examine have to be made, anyway. Rather than accepting the default strategies for curtailing cognitive overload, we could develop our own.

There is another way of thinking about the relationships among investigators exemplified in figure 2 that may mitigate the implied relativism. One can think of science as traditionally involving two types of evidential activities. The more fundamental of these is the use of sensory information. In addition, the rational processes of the mind are used in order to draw inferences from specific perceptions. In other words, the contentions of science are developed from the ground up. When one seeks to justify a statement, one can resort to logical argument or, better yet, appeal to observations made through the sensory modalities. It has been suggested that a third type of evidential activity also plays a role in science, namely, the utilization of all the images, thoughts, and feelings of an investigator, in addition to her formal observations, which are not supposed to be acceptable for the erection of the scientific edifice of information. If loss of consensus becomes apparent in the step from perception to ratiocination, then consensus dissipates completely when the residual contents of an investigator's experience are admitted as a basis for scientific knowledge. It would seem that the further away one gets from the items of the material world, the greater the degeneration into relativism.

Is it possible that there is also a top-down process, that sensory impressions and logical argument are not the only means of justification that can lead to consensus? Perhaps, within the residue of experience traditionally unacceptable to science, there are potential faculties for knowledge that would also allow for agreement among investigators. Moore and I found that 55 percent of respondents agreed with the statement "There are modes of understanding latent within a person which are superior to rational thought." Only 30 percent of respondents disagreed with that statement. Given that this item is a constituent part of the extraordinarily transcendent position and is correlated with other items concerning extraordinary beliefs and inner growth, it is reasonable to think that respondents who agreed with this item believed that there are ways of understanding other than those associated with the sensory modalities and ratiocination. If such a mode of understanding were to exist and be developed, would there be agreement among investigators?

**CHAPTER FOUR**

# Transcendence

## *The Promise of Enlightenment*

> A little learning is a dangerous thing;
> Drink deep, or taste not the Pierian spring:
> There shallow draughts intoxicate the brain,
> And drinking largely sobers us again.
>
> —*Alexander Pope, "An Essay on Criticism"*

Thus far in this book we have been mainly focused on the groping and not on the elephant. We have been circumambulating the existential questions without looking at the answers that have been proposed for them. Toward the ends of both chapters 2 and 3, we considered the possibility that we may possess latent faculties of knowledge superior to rational thought. Could it be that it is possible to transcend the constraints of perception and conation and to know reality in a more direct, profound way? Is transcendence of the human condition possible?

The notion of transcendence has had a troubled past. Arguments have had to be made in the academic literature justifying the right to use the concept in scientific writing concerning religious experiences.[1] Whether or not there is anything in reality that corresponds to the concept is open to investigation. For our purposes, let us start with a meaning of the term that is partially relativized to experiential events: "Transcendence refers to the very highest and most inclusive or holistic levels of human consciousness."[2] Using this definition may remove some of the uneasiness associated with talking about transcendence, although such relief will only be temporary, given the types of experiences that will be discussed later in this chapter. More generally, the term "transcendence" can be used to designate the surpassing of limitations, including those of human experience. Hence it can denote radical changes of being.

In this chapter we will start by looking at some answers to the existential questions. This will lead us naturally into an examination of meditation and transcendence. Toward the end of the chapter, the experiences and philosophy of Franklin Wolff will be discussed in some detail, including some of his ideas concerning a latent faculty of knowledge and the relationship of mathematics to transcendent states of consciousness.

### The Meaning of Life

What have students in my humanistic psychology course said on the first day of classes about the meaning of life and the purpose of their existence? Their responses to these questions can be grouped into five overlapping categories.

The first category consists of metacomments. Rather than answering the questions, students talked about issues involved in answering them. Some said that the questions were silly. Others, that these were things that most people do not think about unless they are confronted with a crisis. Some acknowledged that they did not know the answers but hoped that life had some meaning because the alternative was too frightening. Some said that it was less important to question the meaning of life than it was to enjoy it and to learn from one's experiences. Others said that answers did not come from thinking but from practical experience. Finally, there were those who stated that what was meaningful would be unique to each individual.

I placed specific purposes in a second category. These included bringing fulfillment to one's parents' lives, being a role model and caregiver for one's daughter, teaching, developing a career, and making an impact on the world. For some, belief in God and preparation for that which comes after death were meaningful. In some cases, students have said that the purpose of life was to have as much fun as possible.

In the third category, meaning in life results from our interactions with others. Thus meaning is found in belonging to a group, taking responsibility, and working cooperatively. Each person is unique and gives meaning to others. The purpose of life is to learn how to love.

The fourth category consists of abstract characterizations of the meaning of life. Some students maintained that there was no meaning, that we were just blobs of protoplasm in a meaningless universe. Others characterized meaning in terms of the achievement of subjective states of consciousness, such as happiness and tranquillity or congruence between one's actions and one's beliefs and values. Some said that there was meaning in the attempt to find meaning; and others, that one's purpose was to be, that existence was inherently meaningful.

I placed self-actualization in a fifth category by itself. Some students thought that the purpose of life was to seek to understand and fulfill oneself through knowledge, experience, and sharing. In this case, the meaning of life would be revealed symbolically through the course of one's spiritual development.

Questions concerning the meaning of life have been addressed by Victor Frankl, a Jewish psychiatrist interned in concentration camps during the Second World War. According to him, there are three ways in which a person could discover meaning. It could be realized by "creating a work or doing a deed; . . . by experiencing something or encountering someone; and . . . by the attitude we take toward unavoidable suffering." Satisfaction with life ensues when one transcends oneself through participation in the world. In the first case one is engaged in the work of completing a project. In the second case, truth, beauty, goodness, and love can be experienced through participation in nature, culture, or love for another human being. In the third case, one faces unavoidable suffering with a change of attitude. One transcends oneself by accepting the responsibility of the challenge that life has posed.[3]

In terms of my students' answers given above, Frankl's discussion of existential questions would fall into the first category. The first two of his ways of discovering meaning could be identified with the second and third categories, respectively. The need to change one's attitude so that one lives up to life's challenge would be consistent with the spirit of some of the responses in the fourth category. However, Frankl rejected the notion that an effort toward self-actualization could lead to meaning, so his views would be largely incompatible with the fifth category.[4]

How satisfactory is the relativization of meaning? By allowing answers to the question, "What is the purpose of my existence?" many student responses were concerned with that which is meaningful for specific individuals rather than with a universal answer for the meaning of life. Frankl has also made this distinction by using the term "super-meaning" to refer to the "ultimate meaning of human suffering."[5] Such super-meaning, he contended, exceeds the capacity of the rational mind.

This is a pivotal point. It is meaningful to ask existential questions. "What is the meaning of life?" is a rationally coherent question. But we may not be satisfied with any of the answers that we can come up with. The answers for which we are looking may be rationally indeterminate.[6] It is even less likely that they are empirically available, that we can somehow encounter them through our sensory processes. But then where does that leave us?

From a scientistic point of view, there is no meaning. Life is meaningless, and we need to resign ourselves to "reality." From a religious point of view, we can believe that there is a meaning, even though it may be hidden from us. We could live by faith. There is a third alternative. Perhaps we can understand the meaning of life, but we must first develop the latent faculties necessary to do so. We may not be convinced that such faculties exist. However, one way to try to find out would be to undertake a process of self-transformation that may lead to the development of modes of understanding superior to rational thought.

This is not an obscure idea. "Whereas knowledge is something we have, wisdom is something we must be. Developing it requires self-transformation."[7] This same idea appears in the fifth category of student responses concerning the meaning of life. Meaning is revealed as a by-product of the effort to actualize one's potential. Moore and I found that the extraordinarily transcendent position with regard to beliefs about consciousness and reality was characterized in part by an emphasis on the discovery of meaning through inner exploration and the necessity of self-transformation in order to fully understand consciousness. Self-transformation could be achieved through meditation or a spiritual way of life.

But now we encounter a small dilemma. The development of one's latent potential for its own sake has come under serious criticism. For Frankl, preoccupation with oneself precludes the possibility of existential fulfillment. Rather, it is through self-detached encountering of the world that meaning ensues. Furthermore, in chapter 2, the problems with a self-serving attitude were touched on, including the dehumanization of others who are subordinated to our needs for self-development.

In response to Frankl, Abraham Maslow has addressed these types of concerns. For Maslow, at some point in one's development, dichotomous categorization of one's experiences becomes difficult as self-transcendence and actualization of one's self fuse.[8] For example, if one appreciates another person in her uniqueness, is this other- or self-serving? If one's greatest sense of fulfillment comes from loving others, is this other- or self-serving? Others may simultaneously be respected in their own right and instrumental for one's actualization. The genuine development of one's potential need not end up being solely for one's own sake.

Self-actualization and encountering the world are also inherently two aspects of authentic science. A process of transformation is necessary for a scientist to neutralize the detrimental effects of preconceptions and to adopt an open-minded but discriminating attitude. At the same time, a scientist's project is to deepen her understanding about some aspect of reality. This deepening is again a transformation that may lead to actions predominantly serving others.

Of course, a scientist could seek to deepen her understanding of the meaning of life. Conversely, since up to a point we all naturally use the same strategies as scientists, developing our understanding of the meaning of life could be regarded as a scientific activity.

**Meditation**

Existential questions are rationally coherent. However, satisfactory answers may not be available through empirical or rational means. The effort, then, becomes one of trying to determine if there is a latent mode of understanding superior to rational thought and, if it exists, to bring it into activity. How does one go about doing this? In this section let us develop one line of speculation about this effort.

While it is the discursive aspect of the mind that allows for explicit formulations of existential questions, at the same time it is that aspect of the mind that prevents the surfacing of superior modes of understanding. The normal functioning of the mind obliterates the transcendent. "The mind is the great slayer of the Real. . . . Let the disciple slay the slayer."[9] The task thus becomes one of eliminating the barrier created by the discursive mind. This can be done in one of two ways, through regression or transcendence.

In order to put the mind in its place, one can seek to prevent its development in the first place or, if it has partially developed, to suppress its functioning in the hope that it will atrophy. In that case, the active ingredients in attempting to uncover the transcendent might be good deeds and the expression of intense emotions such as devotion and love. The mind becomes negatively valued. The discriminative faculty remains undeveloped, so that only simple schemata are used for interpreting experience. For this method to succeed, a specific set of beliefs must be chosen, or acquired through circumstance, that cannot be questioned. Discordant ideas are blocked out, for example, by preventing exposure to them or labeling them as satanic. Because of the suppression of discursive thought, those who follow this path are unlikely to describe it in these terms.

Alternatively, one can seek to transcend the functioning of the mind. Techniques that are used for this purpose are referred to as "meditation." These can be divided into three broad categories, depending upon the manner in which attention is directed.[10] In witnessing meditation, one observes the ongoing activity of the mind; in concentrative meditation, one restricts attention to specific mental contents; and in reflexive meditation, one seeks the source of consciousness. In each case, the instructions for meditation are simple, the difficulty lying in their application.

Before describing in greater detail the techniques that fall within these three categories, it should be noted that there is a bifurcation of the academic literature concerning meditation. Only some of it has pertained to meditation for the purposes of transcendent realization; a large part has been about the use of meditation for relaxation. Because meditation has been used as a means of relaxation and because of its prominence in the academic literature,[11] let us start with a discussion of some of the aspects of transcendental meditation.

Transcendental meditation was formulated as a derivative of Shankara's eighth-century Hindu teachings by Maharishi Mahesh Yogi and popularized after his arrival in California from India in 1959.[12] To do transcendental meditation, one sits comfortably with eyes closed for about twenty minutes where one will not be disturbed and silently repeats a word over and over for that period of time. When thoughts stray from the word, one recognizes that thoughts have strayed and brings one's attention back to the word. That is it.[13]

When one is initiated into transcendental meditation, one is given a "mantra," a word that one is to use and never to reveal to anyone else. Mantras are supposedly assigned so as to match the unique characteristics of the meditator. However, the only characteristics of meditators that appear to have been taken into account have been their age and sex. Subsequently, sex distinctions seem to have been dropped and mantras appear to have been assigned on the basis of age group. There has been a list of sixteen such mantras, whose Sanskrit meanings are associated with Hindu divinities, such as Lord Krishna and the Divine Mother.[14]

Herbert Benson, one of the first investigators to study the physiological changes associated with transcendental meditation, found that during the time that someone is meditating, there are decreases in oxygen consumption and carbon dioxide elimination, lowered heart and respiratory rates, decreased blood lactate levels,[15] and increased alpha and theta waves as measured with an electro-encephalogram. These transient changes were called the "relaxation response" and are consistent with decreased sympathetic nervous-system activity.

Benson found, however, that transcendental meditation was not the only way in which the relaxation response could be induced. One could also use autogenic training, which involves the imagination of bodily sensations; progressive relaxation, in which one systematically decreases tension in skeletal muscle groups; and hypnosis, which increases one's vulnerability to suggestion. Benson has also developed a "simple secular technique" that can elicit the relaxation response. The word "one" plays the role of the mantra in this technique, which is otherwise similar to transcendental meditation.[16]

Other investigators have also studied the changes attributed to transcendental meditation. There is some agreement that the use of transcendental meditation and similar techniques can result in lowered anxiety,[17] although in some cases anxiety has become elevated to pathological levels.[18] Stress-reducing effects may be due to sleep, since some meditators appear to spend much of their meditation time asleep as determined by electroencephalograms and electro-oculograms.[19] In fact, the beneficial physiological effects of transcendental meditation do not appear to differ substantially from those produced by resting.[20] Ironically, when subjects practiced "an antimeditation technique"[21] for which they were instructed, in part, to "engage in deliberate cognitive effort that is directed in a positive direction," the same degree of stress reduction was found as when subjects practiced a simpler, more passive cognitive technique similar to transcendental meditation.[22] In other words, there are many different ways in which the relaxation associated with transcendental meditation can be produced.

To a great degree, there has been an identification of meditation with decreased physiological arousal.[23] However, not all types of meditation have been designed to produce relaxation. Indian practitioners of kriya yoga, a specific meditation technique, have exhibited accelerated heart rates as well as high-amplitude, high-frequency beta-wave activity as measured by an electroencephalogram.[24] To some extent the actual physiological changes associated with meditation appear to vary with the technique used and the intentions of the practitioner.

We have been focusing on stress reduction and relaxation because meditation has largely been studied in terms of its practical benefits. Various meditative techniques have been used in clinical and military settings in order to try to enhance psychological and physical functioning.[25] Since our purpose is to discuss meditation as a means of transcending the discursive activity of the mind, let us turn now to descriptions of witnessing, concentrative, and reflexive types of meditation.

Witnessing styles of meditation, often associated with the Buddhist religion, involve observation of one's normal stream of consciousness. One sits with back erect and eyes open and pays attention to the sensations, feelings, and thoughts that occur within one's awareness. As these events occur within the domain of the mind, one labels and dismisses them without making judgments. If judgments are made, these are simply regarded as additional events to be labeled and dismissed. Initially, thoughts may wander. When one realizes that thoughts have wandered, they are labeled and dismissed. Mindfulness is the process of attending, labeling, and dismissing. As this process is sustained, mindfulness develops into insight, whereby realizations about the nature of mind emerge. There are variations, such as walking instead of sitting, registering rather than labeling, and restricting the locus of attention.[26] However, the idea is to witness the ongoing activity of the mind in order to understand its nature.

Even though Freud was apparently unaware of the Buddhist parallels, witnessing meditation is similar to the discipline of evenly suspended attention on the part of a psychoanalytic therapist. In the traditional practice of psychoanalysis, a client would be asked to communicate uncritically all that occurs to her, while the analyst would "suspend . . . judgement and give . . . impartial attention to everything there is to observe." In this manner, the analyst could "catch the drift of the patient's unconscious with his own unconscious." The development of the skill of "equal attention to changing objects of awareness . . . is the stated purpose of most Buddhist meditation practices."[27]

Concentrative styles of meditation have been associated with both the Hindu and the Buddhist religions.[28] Piero Ferrucci, writing within the tradition of psychosynthesis, has used the metaphor of public and private locations in order to illustrate the process involved. A town square is a public place to which everyone has access, whereas a home is a private place that only certain individuals can enter. Analogously, during the time of meditation, one changes one's mind from a place to which thoughts have unrestricted access to a place in which some thoughts are allowed but not others.[29] This can be done by selecting an object of meditation—either a seed thought or symbolic image[30]—about which one thinks. The theory is that the discursive mind gets bored when its range of activity is confined and becomes still, thereby allowing the transcendent to emerge in consciousness, which reveals the deeper meanings of the object of meditation. While there may be a natural inclination to simply try to stop thinking, there have been strong warnings against forcing thoughts to a stop.[31] One is to sustain attention without strain.

Is concentrative meditation so different from what a scholar or a scientist does in order to solve a problem? She focuses her thoughts on the object of interest, brings to bear all pertinent information, and confines her thoughts to those which are relevant in order to understand the problem more deeply. Is it also possible that she exhausts the resources of the discursive mind and draws on a latent faculty?

Graham Wallas has described four stages in the process of problem solving. The first stage, preparation, involves the application of one's knowledge and skills to the investigation of the puzzle. In the second stage, incubation, one abstains from consciously thinking about the problem, allowing unconscious exploration to run its course. The third stage, illumination, is characterized by insight concerning the resolution of the enigma. Verification of the solution is carried out in the fourth stage using the cognitive and vocational resources at one's disposal.[32] While this may not be an accurate portrayal of the actual cognitive processes involved in problem solving, it does give a description, from an experiential point of view, of the way in which scientific and scholarly activity may already involve the transcendent.

The third style of meditation could be called "reflexive meditation." Rather than noticing the contents of one's mind or focusing on a given object of consciousness, the meditator seeks to find the subject for whom there are contents. Right away, however, the descriptor "reflexive" has to be clarified and the description of the task qualified. One is not to seek the source of consciousness as an additional target of attention, as the term "reflexive" would imply. Anything that can be conceptualized as an object of awareness is not that which one seeks. However, while it is easy to characterize the task in terms of what is not involved, it is difficult to indicate what it is that is involved, since, presumably, that can only be understood using the internal processes whose activation one is seeking. The nature of this type of meditation, as well as some of its consequences, will be further characterized in our discussion concerning Wolff's experiences and philosophy.

Regardless of the style of meditation one is practicing, one is supposed to remain alert. Whether one is watching the contents of one's mind, focusing on specific objects of meditation, or seeking the source of consciousness, meditation is not an activity in which one simply "spaces out," allowing the mind to drift without purpose.

From the point of view of cognitive science, there are no latent faculties to be uncovered. Reality is ontologically characterized by a pervasive objectivism that allows one to ascertain facts through sensory and rational processes. As Moore and I found in our empirical study of beliefs about consciousness and reality, respondents tending toward a materialist position tend also to consider consciousness always to be consciousness of or about something.

Allusions to consciousness without an object are reformulated in terms of that which is objective, so that such aspects of consciousness are regarded as instances in which consciousness is applied to itself.[33] More generally, from the materialistic point of view of cognitive science, introspected objects of attention are often considered to be confabulated, largely misleading representations of one's cognitive processes.[34]

Later in this chapter, we will look at examples of the use of the three styles of meditation. First, however, let us consider experiences of transcendence more generally.

### Enlightenment

Around the time that I was twenty years old, I had a glimpse of the transcendent states of consciousness that I had previously read about. I had gone over to visit two friends who were planning a trip to northern Africa. We were in the kitchen, talking. I was sitting, by myself, on the floor looking at a large map of the Sahara Desert, which was spread out in front of me. My friends were talking about an oasis in the Sahara Desert that they were hoping to visit. As one of them came over to show me the location of the oasis on the map, I entered a satisfying state of being that continued for some minutes. During this time, the interaction with my friends continued unimpeded, so that I continued to function in an ordinary way. Subsequently there were no apparent residual effects.

There was a noetic aspect to this experience. I felt that I knew that I had reached a greater state of lucidity than that present during my normal state of consciousness. It was also a synthetic state in that I still knew all the things that I ordinarily would know. Not only were all my ordinary thoughts fully intact; so were all my problems and frustrations. However, during this time, I could see how all my difficulties were part of a harmonious order within which they fit perfectly. In fact, my trying experiences were part of a process of which the inevitable end result was the manifestation of perfection. It seemed to me that the only question was the length of time that it would take for this state to be realized. The more I could relinquish anything that was imperfect, the sooner it could occur.

During this experience, my perception was that my understanding was greatly expanded. I realized that my ordinary thoughts and experiences were incomplete fragments fitting naturally into a perfect reality and that this was what reality was really like.

There was also an affective aspect to this experience. I felt a complete and pervasive sense of well-being. While this joy was refined in quality, it was also intimately sensuous and immediate. Despite that, there was nothing frantic about it. Rather, the joy was accompanied by a deep sense of peace. I cannot imagine a more desirable state of emotional well-being.

Religious and mystical experiences have been studied in psychology, notably within the psychology of religion and more recently within transpersonal psychology. My experience has some of the prototypical features of a mystical experience as it is usually understood. Thus, this experience was a "valid source of knowledge," it was characterized by positive affect, and unity was perceived in the diversity of all things. Mystical experiences can also have features that were not prominent in my experience. They can be inherently inexplicable. They can be characterized by a sense of "holiness" that is not necessarily allied to any particular theology. They can defy traditional logic. All things may seem alive. One may experience timelessness and spacelessness. And one may experience consciousness devoid of any content, in which the sense of self has also dissolved.[35]

When surveys have been conducted to determine the frequency of mystical experiences in the general population, more than one-third of all people claim to have had such experiences. Bernard Spilka, Ralph Hood, and Richard Gorsuch have reported that 35 percent of their subjects indicated "a positive response to questions regarding the occurrence of mystical experiences."[36] The Gallup Organization has found that 43 percent of people in the United States have "had an unusual spiritual experience."[37] Moore and I found that 47 percent of respondents indicated that they had "had an experience which could best be described as a transcendent or mystical experience."[38]

One has to be careful, however, in interpreting these figures. Eugene Thomas and Pamela Cooper found that 34 percent of their sample of 305 individuals claimed to have been "close to a powerful spiritual force that seemed to lift [them] outside of [themselves]." However, their more detailed inquiry revealed that only 1 percent of their sample had had an experience involving overwhelming emotions or "sense of the ineffable, feeling of oneness with God, nature, or the universe." Eight percent had had psychic or unusual experiences, and 16 percent had had experiences of faith or consolation which involved religious or spiritual events without paranormal elements and usually within the context of a traditional religious system. The remaining approximately 10 percent had given irrelevant or uncodable responses.[39]

The results of the study by Thomas and Cooper indicate the necessity for distinguishing between experiences that are simply unusual and those that are deeper. There are different characterizations of this difference. For example, "psychic" refers to paranormal experiences, while "spiritual" refers to those that have existential value, whether paranormal features are present or not. Similarly, the terms "extrapersonal" and "transpersonal" have been used to distinguish these types of experiences, while Assagioli has used the words "horizontal" and "vertical."[40] Problems arise when the experiences along the horizontal dimension are assumed to have the attributes of those along the vertical.

Mystical experiences have often been considered to be pathological.[41] However, in a study based on the responses of 285 psychotherapists, it was found that "most therapists did not view such experiences as necessarily pathological." Four and one-half percent of respondents' clients had reported having had a mystical experience within the previous twelve months, and 50 percent of the responding psychotherapists "reported having had a mystical experience at some time in their lives."[42] In other studies, efforts to find correlations with indicators of psychopathology have produced negative results.[43] On the contrary, there is some evidence that those who claim to have had transcendent experiences are more emotionally stable, more open-minded, and more tolerant of ambiguity[44] than those who do not claim to have had such experiences. Similarly, having had "religious experiences in prayer" is correlated with life satisfaction, existential well-being, and happiness.[45] Rather than being pathological, transcendent experiences are consistent with psychological well-being.[46]

Are there predisposing factors for mystical experiences? There is the notion that one can prepare oneself for, but cannot command, a mystical occurrence. However, summarizing empirical studies concerning triggers for mystical experiences, Spilka, Hood, and Gorsuch found that anything that "suddenly emerges or is recognized to point to the limits of everyday reality may suddenly serve as a trigger to mysticism."[47] Perhaps one transcends one's ordinary functioning when the limits in one's life become apparent.

On the basis of empirical evidence, Peter Nelson has developed a three-dimensional model organizing the correlations of variables with occurrences of "praeternatural experiences," a broad category that has included mystical experiences as well as "remote perceptions" and "shamanistic-like 'other-world' adventures."[48] He has divided preternatural experiences into those that are predominantly perceptual and those that "give rise to a sense of ontological otherness."[49] These two categories would approximately correspond to the horizontal and vertical types of experiences described above. In this model, phenomenological features, operational factors, and personality variables are correlated with normal, perceptual, and ontic experiences.[50] Thus, ontic experiences are phenomenologically characterized by ontological reorientation, intense affect, sense of salvation, and transcendental quality. They are activated by stress, immersion in an activity requiring concentration, social isolation, abrupt alteration of affective state, abstention from enjoyable activities, and lack of control.[51] Of the personality measures that he used, Nelson found that the best predictor of frequent preternatural experiences is absorption,[52] the capacity that a person has for becoming preoccupied with the object of her attention. Absorption is correlated with hypnotizability, the purported occurrence of extrasensory perception and out-of-body experiences,[53] and long-term practice of meditation.[54] In addition to absorption, personality characteristics associated with ontic experiences include higher levels of positive affect and lower levels of aggression than those associated with non-preternatural experiences.[55]

There are a number of ways in which attempts have been made to discount the noetic value of transcendent experiences. For example, given that people who score higher on absorption are also more likely to have had more mystical experiences, could it be that these experiences are just the product of an overactive imagination? Perhaps one becomes so involved in transcendent ideas that one loses the power of discrimination and comes to regard them as real.

Based on a comprehensive review of the relevant neurophysiological literature, mostly concerning the effects of psychotropic drugs, Arnold Mandell has given an account of transcendence in physiological terms. Drugs such as hallucinogens, amphetamines, and cocaine disturb the activity of the neurotransmitter serotonin in such a way that it fails to inhibit excitation of hippocampal $CA_3$ cells—a sector of cells in the hippocampus of the brain. Such an effect would also be created by prolonged activity, stress, sensory deprivation, or meditation. Because of this hyperactivity of the hippocampal $CA_3$ cells, the hippocampus cannot properly compare information coming from the outside with information that is already stored in the temporal lobes. In effect, internal information is unchallenged by external data, leading to a sense of unity with the universe and the conviction that one already knows everything. Hippocampal $CA_3$ cell hyperexcitability leads to "hippocampal-septal synchronous discharges," which are experienced as emotional flooding and may be variously interpreted—for example, as ecstasy. These cells become "progressively excited and die," resulting in permanent personality changes, such as "religious conversions, hyposexuality, transcendent consciousness, good nature, and emotional deepening," which are characteristic of psychomotor epileptics with hippocampal cell death. Mandell has speculated that in the absence of external explanations, the source of the experience of ecstasy, as $CA_3$ cells fire furiously, followed by "empathic beatitude" once they are dead, gets erroneously attributed to God.[56]

Does this mean that transcendent experiences are nothing more than the hyperactivity and death of brain cells? An analogy may be helpful for putting this information into perspective. If I am watching a hockey game on television and somebody comes along and puts a sledgehammer through the screen, this does not mean that the hockey game has ceased to exist. In other words, the existence of physiological correlates of some experiential events does not rule out the possibility of experiences at another level of reality. Death of hippocampal cells may be the source of transcendent experiences, or it may simply be one possible physiological mechanism through which such states of consciousness can manifest themselves. Physiological explanations cannot discount the possibility of transcendent consciousness.

Another way of attempting to discount transcendent experiences is to suppose that the contents of one's absorbed state are a product of one's culture. From a constructivist point of view, "mystical experience is significantly shaped and formed by the subject's beliefs, concepts, and expectations"; the implication is that there is nothing genuinely transcendental about it. On the other hand, those aligned with the "perennial philosophy" would maintain that there are transcultural, core characteristics of mystical experiences that get interpreted according to the categories, beliefs, and language that are brought to them. It is assumed that these core experiences are not just common physiological effects but result from "direct contact with a (variously defined) absolute principle."[57]

While philosophical debate may clarify some of the issues involved, it cannot resolve the question of whether or not there really is anything more. Of course, having perhaps experienced and become convinced of the existence of something more, one may have lost essential brain cells and be in no position to pass judgment on the validity of one's experiences. On the other hand, the scientist abstaining from the benefits of direct experience may be lacking brain changes that would allow her to understand the true nature of the matter.

One belief, sometimes associated with the perennial philosophy, is the developmental thesis that human beings can evolve not just to their present state of achievement but onward, to exceptional states of well-being in which transcendent experiences occur with increasing frequency.[58] There is some debate as to the extent to which one takes one's physical body along on this voyage of self-transformation.[59] If the developmental thesis were true, then it would mean that transcendent experiences would be possible only after a person has surpassed the limits of normal human development. In particular, they should not occur in childhood. However, there is empirical evidence indicating that children can have genuinely transcendent experiences.[60] Any of the features of the adult experiences appear to have the potential of being encountered also by children. For example, they can see meaningful visions, experience another level of being underlying nature, and become aware of the unity of all of creation. The developmental thesis can be salvaged by reinterpreting it, not with reference to the personality but rather with reference to the degree of integration of the personality with the transpersonal self. Thus, while an "old Soul"[61] may need to recapitulate normal childhood development during any given lifetime, transpersonal irruptions are nonetheless possible. However, this presupposes that reincarnation, or some form of transmigration of the soul, actually does occur.[62]

Whether or not the perennialist position is correct, transcendent experiences usually have a great impact on those who have had them. One of the phenomenological characteristics of ontic experiences as given by Nelson is that of ontological reorientation. This impact becomes apparent when reading specific cases of such experiences.[63] Moore and I found that people who claimed to have had transcendent or mystical experiences were also likely to claim to have dramatically changed their ideas about life.[64] Transcendent experiences bear great likeness to near-death experiences. In fact, Kenneth Ring has argued that "in their essence, *NDEs have nothing inherently to do with death at all*" but are opportunities to realize the limitations of any personal expression of the self and attain "spiritual immortality."[65] Perhaps the impact that near-death experiences have on an experient's subsequent attitudes is due to the same noetic features that change those who have had transcendent experiences.

**Examples of Meditative Experiences**

Meditation is supposed to facilitate the self-transformation necessary to understand the answers to existential questions, but it is rarely practiced in isolation from other life-style changes. Thus when transcendent or other unexpected experiences do occur, it is difficult to know what to attribute them to. Nonetheless, examining the experiences of those who have meditated will provide some insight into the meditative process. In this section, I shall discuss the experiences of Roger Walsh and Douglas Baker, who used witnessing and concentrative styles of meditation, respectively. Those of Wolff, who used a reflexive style of meditation, will be presented in the next section.

As part of his postdoctoral training in psychiatry, Walsh had to undergo a process of psychotherapy. He did this with a psychotherapist of humanistic-existential orientation, who would "train his clients in the development of heightened sensitivity to, and appreciation of, their own inner experience and subjective world." Within weeks, Walsh found an increase in his "subjective sensitivity." He began to experience synesthesia, whereby he would perceive stimuli in sensory modalities other than the ones in which those stimuli were presented, so that, for example, he would see and feel sounds. Furthermore, this increased awareness allowed him to identify "a constant flux of visual images" and to recognize that "these images exquisitely symbolized what I was feeling and experiencing in each moment."[66] This access to the inner world proved to be an unexpected source of personal information. He came to see that he was the intentional creator of both his pleasurable and painful experiences and that he had to acknowledge responsibility for them.

Walsh has said that his beliefs started to change. He realized that his assumptions acted as self-fulfilling prophecies, placing unnecessary constraints on him. He also saw that many of the ideas he had about his own nature, which were consistent with the cultural belief system, were incorrect. For example, he found that he could function periodically without experiencing anxiety and guilt, even though these negative affective states are considered essential for normality and their absence is considered a symptom of psychopathy. He thought that he had also been wrong in his previous negative assessment of the "human potential movement and various growth-oriented trainings," which he subsequently explored in a more open-minded manner.

Walsh learned a witnessing style of meditation whereby he sought to "maintain continuous awareness and to allow attention to focus on whatever stimulus is predominant at any time" while he was meditating. The more sensitive his meditation became, the more he realized that his mind was filled with fragmented "thoughts and fantasies" rather than the purposive rational processes that he had assumed were there. He found that these interior preoccupations were harder to control in meditation than they had been during psychotherapy because of their greater intensity and the lack of modulation by a psychotherapist. Walsh found that the contents of mind that disturbed his concentration were those to which he was attached and "around which there was considerable affective charge." For example, anxious thoughts would arise that were the result of his attachment to a "scanning process" for the detection of fearful stimuli. Not only would the scanning process create an alarm over anything that resembled fear; the thought of eliminating the scanning process itself produced fear at the prospect of not having anything to do in its absence.

Another experience that Walsh attributed to his meditation was the inability to "know what anything [meant]" for some time after meditating. He found that the familiarity of the world was dependent upon the categories used for labeling sensory impressions. Once this labeling process is eliminated, if only temporarily, the world feels less familiar. It became apparent to Walsh that we live in response to the labels that we use rather than to the events themselves as they are. We live in a staggering state of ignorance, sharing "a mass cultural hypnosis and psychosis." "Indeed . . . culture can be viewed as a vast conspiracy against self-knowledge and awakening in which we collude together to reinforce one another's defenses and insanity." Walsh perceived himself to be on a journey, so that, as his meditation deepened, he expected that there would be other unexpected experiences.

In Baker's description of concentrative meditation, four stages are involved. In the first stage, one prepares for meditation. Such preparation includes the establishment of a physical space conducive to meditation, practicing of eye and breathing exercises, keeping of a spiritual diary, and seating oneself so that one's spine is erect. Withdrawal of attention from sense impressions constitutes the second stage. The third stage consists in concentration, whereby attention is focused in a relaxed manner on a single object projected "on to the screen of the stilled mind." The three factors of breathing, concentration, and visualization of the object must combine in order for the fourth stage, that of meditation proper, to be successful. In this stage, one becomes identified with the object and with the "root causes behind all form." Occasionally, with the body and mind still, consciousness "bursts through into another world."[67]

Baker has described experiences that have occurred to him that would be characteristic of this final stage of meditation. After six months of reading theosophical literature, following a period of meditation, his body was jerked into a rigid state and he felt as though a force were being thrust through him. There were three main features associated with this experience. He had the sense of "looking through into a new world even though fully conscious." This perception transformed, so that he had an experience of interior light. The perceptual experience was accompanied by feelings of ecstasy and a "sense of eternity." During the subsequent decades, this experience would return four to forty times every week.[68] Baker has interpreted these experiences as the downpouring of fire, whose control must be mastered so that it can be applied to the service of humanity.[69]

### Wolff's Introceptualism

Franklin Fowler Wolff was born in 1887.[70] After four years at Stanford University, Wolff graduated in 1911 with a bachelor's degree in mathematics, with philosophy and psychology as minor subjects. After one year of graduate studies in philosophy at Harvard University and a subsequent year of teaching mathematics at Stanford University, he relinquished a promising academic career in order to fully pursue the development of his spiritual interests. While Wolff drew on numerous sources for his understanding of reality, his interest in theosophical teachings and study of the philosophy of Shankara had significant and pervasive influences on him. His efforts were rewarded with two fundamental realizations—on August 7, 1936, and September 8, 1936—the effects of which persisted until his death in 1985. For 101 days following his first realization, he kept a journal, which was published as *Pathways Through to Space.* Subsequently he wrote *The Philosophy of Consciousness without an Object,* in which he has described his experiences and situated them within the context of Western philosophy.

In addition to the context provided by traditional Western academic training, Wolff emphasized clarity of understanding and an effort to communicate the substance of his realizations without recourse to religious language. He was a man of great personal integrity who felt strongly that one should use words for the "maximum possible correctness" with respect to their "truth value" rather than using them for their psychological effects as instruments of influence.[71] For these reasons his descriptions of transcendent experiences are particularly compatible with a scientific approach.

Wolff had become convinced of the value of seeking a "transcendent mode of consciousness that could not be comprehended within the limits of our ordinary forms of knowledge." From the time of this conviction, as a graduate student during 1912–13, until the occurrence of the first fundamental realization in 1936, Wolff had a number of key experiences that contributed to the liberation. Most important of these were insights that allowed him to develop an understanding of the relationship between ordinary and transcendent states of consciousness as well as an effective means of realizing the transcendent states. According to Wolff, some of these insights were not just a matter of intellectual knowledge but involved an additional quality of "value" that gave them depth that they otherwise would not have had. This depth is "precisely the inexpressible element in all Gnostic and Mystical Realization" that cannot be defined but whose "actuality is indubitable" when it is present.[72]

Wolff realized that ordinary consciousness is characterized by a duality between subject and object, while in the case of transcendent consciousness that duality is "destroyed." The common element for both relative and transcendent consciousness lies in the "subject or self."[73] This subject can never itself be an object of consciousness; and because it stands outside of time, it is immortal. Through a process of discrimination, Wolff recognized himself to be that self.

Another insight, which proved to be the "crucial key to the transformation process," was that "substantiality is inversely proportional to ponderability." That which is real, substance or depth, is missing precisely to the degree that something is experienced or conceptualized. "Fullness [was found] in just those zones where sensation and conception reported absence of anything."[74] At this point it remained only to retract one's self-identification from the phenomenal world of experience and to identify with the substantial. This was accomplished by a process of meditation in which Wolff sought to isolate the "subjective pole of consciousness," placing his "focus of consciousness . . . upon this aspect." No effort was made to interfere with the ongoing processes of thought or to seek a subtle object of consciousness or experience. Nor was there any expectation of a "new experiential content in consciousness."[75]

To distinguish between events that occur within the relative and transcendent domains of consciousness, Wolff has used the terms "experience" and "imperience," respectively. On August 7, 1936, his first fundamental realization occurred, in which Wolff had the imperience of being "consciousness . . . dissociated from the object" and identified with the self, which was nothing. This was, however, only a "transitory" state in that he realized that the voidness was substantiality, meaning, and superior value. These were some of the characteristics of "consciousness-without-an-object-but-with-the-Subject."[76] Precisely because he had transcended the relative domain, however, language, which functions within that domain, cannot adequately characterize the imperience. Transcendent consciousness is necessarily ineffable. However, although Wolff did not anticipate this, some effects of this imperience precipitated into the relative domain as experiences that could be described.

These experiences had largely cognitive and affective aspects. Thus Wolff's self-concept changed permanently from a space- and time-bound self to a self that was immortal and "[spread] out into an unlimited 'thickness.'" This was accompanied by a sense of freedom, freedom from guilt and resolution of the "wrongness" inherent in life, and by feelings of serenity, joy, and benevolence. There was a depth of knowledge far more abstract than relative thought and a mode of cognition, "Knowledge though Identity," that can "give the fundamental propositions . . . from which systems can grow at once by pure deductive process." The role of information changed, so that whereas Wolff acquired information in the course of the search for the real, it was subsequently used largely to give expression to "abstract and superconceptual" knowledge. The first fundamental realization was not accompanied by any automatisms or trance.[77]

While Wolff found himself in a state of "full enjoyment," so that there was nothing further to be sought, "there was a blot on the contentment" resulting from his awareness of the continued suffering of much of humanity. In this state, his conviction was that essentially all was well and that the apparent deep "wrongness" in the world was "an illusion or a misplacement of our understanding." Nonetheless, he renounced the "great liberation" in order to assist in the "resolution of the problems of suffering or of ignorance that existed in the world." However, he found that this "great renunciation" was a "grievously painful step."[78]

During the night of September 8, 1936, Wolff's second fundamental realization occurred. He again had the imperience of thorough satisfaction. However, this time it transformed itself into a state of high indifference that was characterized by a complete equilibrium between opposites. He found that he could look upon that which is positively valued and that which is negatively valued with "complete dispassion." In particular, the relative universe, with its ignorance and suffering, constitutes one pole of a dichotomy, while the state of satisfaction reached in the first fundamental realization constitutes its opposite. The state of high indifference is "consciousness-without-an-object-and-without-a-subject," from which both the subjective and objective dimensions of reality arise.[79]

While the first fundamental realization had predominantly cognitive and affective features, the second fundamental realization was characterized by power. Thus, from this point of balance between all opposites, Wolff felt that he could act with equal-mindedness and, by an act of will, "enter into delight or into the suffering of creatures with equal [facility]." This was not to say that he was personally apathetic toward the well-being of others or that he did not personally prefer benevolence to malevolence. Wolff himself had chosen the path of righteousness and compassion and advocated that others also choose "the path of exacting moral discipline." While "from the standpoint of the High Indifference" there was "neither merit nor demerit in this," Wolff "chose to continue with the job." Subsequently he found that having chosen to serve humanity was not without compensation and that his life was characterized by "an inner contentment and an inner peace."[80]

How did Wolff know what had happened within himself? We ordinarily accept the functioning of psychical mechanisms that give us knowledge derived from sensory experience and mechanisms that allow for knowledge based on logical reasoning. Wolff has maintained that there is a third "psychical function," which he called "introception" and defined as "the power whereby the Light of consciousness turns upon itself towards its source."[81] It is this latent introceptive faculty that needs to be awakened in order to know transcendent states of consciousness.

Everyday cognitive functioning is characterized by a duality involving an object and a subject, for whom the object is known. When Wolff uses the term "introspection," he refers to the examination of one's experience in such a way that this subject-object duality is retained. In particular, one could objectify the "I" as a construct. But this would not be the "I" that is sought in pursuing the source of consciousness. The "I" that one seeks can never itself be an object of consciousness. The faculty that one seeks to activate is not introspection in this sense. Thus Wolff coined the term "introception" to distinguish the desired faculty from the functioning of introspection within the relative domain.

But how is this latent faculty to be awakened?[82] This is a difficult problem, particularly since its resolution involves not only the consciously directed activity of an individual but also the spontaneous participation of "something other." For Wolff, this faculty was awakened in the process of transformation. The key to the awakening is the mental realization that ponderability is inversely proportional to substantiality. Wolff realized that that which one experiences lies on the surface of life, so that truth can never be found through the activity of sensation and ordinary cognition alone. Rather, in order to awaken the dormant faculty, one must completely renounce everything within experience and surrender to truth.

Wolff has used the analogy of a separation of the stream of consciousness to describe the resultant activation of introception. Ordinarily, consciousness flows as a stream from the subject toward the ponderable universe. However, Wolff found that a separation of the stream was created, with part of it redirected back toward the subject. As this happened, "objective consciousness was dimmed," so that objects lost their "relevancy," while "the consciousness in the state of the reverse flow toward the subject was like a Light highly intensified." The immediate effect was that of being "conscious as the pure 'I,'" which was "the pure Light." However, the union of consciousness "only with the subject" was "transitory," since "almost immediately consciousness [acquired] a new kind of content." Knowledge that was made possible through attention to objects of consciousness within the relative domain was replaced by knowledge through identification with that which is known. "The introceptive realization is a state wherein the subject and the object become so far interblended that the self is identical with its knowledge." One knows because one is that which is known in a seamless domain undivided by categories.

There are various degrees of thinking between the purely relative and the purely transcendental, metaphorically perhaps corresponding to the proportion of the stream of consciousness that flows in each of the two directions. Wolff has called "transcriptive thought" that thought which has elements of both conception and introception. "It may be said that this thought thinks itself, or tends to do so, depending upon the degree of its purity."[83] In this manner whole systems of ideas can be deductively developed from a small number of aphorisms that appear as concepts but are rooted in introception. In particular, whole areas of mathematics can be developed using deductive logic from axioms that are spontaneous productions of an illumined state.

### Mathematics of Transcendence

Wolff stated that his path toward realization of transcendent states of consciousness was that of mathematics, philosophy, and yoga. Mathematics, because of its abstract nature, while still part of the relative domain, is nonetheless close to the transcendent. However, by itself it is sterile; philosophy is necessary to unearth the meaning of mathematics. But to succeed in realizing the transcendent state, mathematics and philosophy as such are not enough. One must also adequately prepare oneself and approach the disciplines of mathematics and philosophy in such a way as to bring about a transformation of consciousness. The proper blend of these three activities defines mathematical yoga.[84]

There are three aspects of the relevance of mathematics to transcendence. First, there are some relationships concerning the nature of consciousness and reality that can be formulated in mathematical terms. Second, more generally, mathematics can provide metaphors for understanding transcendent experiences that are difficult to characterize otherwise. Third, there is the question of whether there are ways in which mathematics can lead to transcendence.

Wolff's key mental realization that substantiality is inversely proportional to ponderability can readily be formulated in mathematical terms. According to Wolff, this is the inverse relationship described by the hyperbola *Substantiality = 1/Ponderability*, as shown in figure 3.

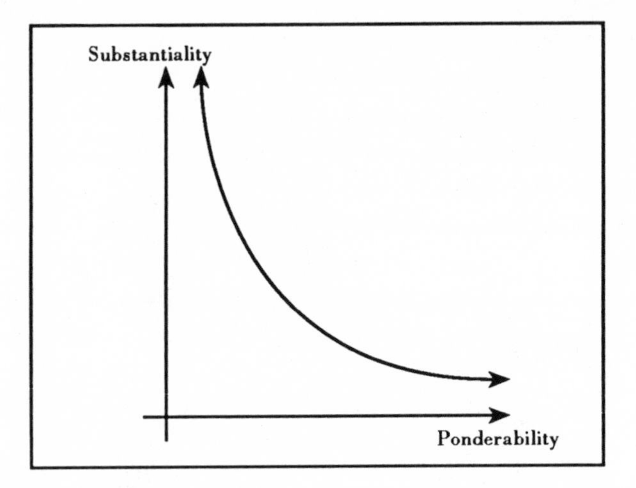


FIG. 3. Substantiality Is Inversely Proportional to Ponderability

The further one goes to the right along the curve, the more concrete the concepts or percepts, and the less there is of substance present. On the other hand, the further one goes to the left along the curve, the more abstract the nature of one's cognition, but the greater is the level of substantiality—that is to say, the closer one is to reality.[85]

In positing this relationship, Wolff has claimed to have resolved an apparent contradiction in the mystical literature concerning the ontological status of everyday experience. It is neither ultimately illusory nor ultimately real but a mixture of both, given by this inverse relationship. More generally, Wolff has maintained that rather than regarding as contradictory the statements of those who have realized transcendent states of consciousness, one must seek to reconcile them. Characterizations of the transcendent within the relative domain must always appear to be contradictory if one is to strive to give a complete description of reality, because the relative cannot contain the transcendent. At different times, emphasis may be placed on one or another feature for didactic purposes.[86]

This same process of reconciliation needs to be applied to Wolff's mathematical descriptions of the relationship of the relative to the transcendent. The hyperbolic function *Substantiality = 1/Ponderability*, according to Wolff, denotes the important principle that progress in attaining the transcendent state of consciousness always involves "inversion of consciousness."[87] Inversion as a hyperbolic function is a continuous curve.[88]

But this inversion is also a discontinuity, as Wolff explained on the day before he described inversion in terms of the inverse relationship of substantiality to ponderability:

> I have passed up and down, as it were, between the relative state and this deeper state, and one thing becomes evident. At some point there's a shift which you instinctively call an inversion of consciousness. . . . At the point of inversion there seems to be something that is akin to what we would call a discontinuity in mathematics where one consciousness blacks out, and immediately another consciousness takes over. Now, there are times when I have deliberately passed up and down, trying to maintain continuity of consciousness here, and it couldn't be done. There was that discontinuity, very quick. On one side, I am, and I state the attitude of the ordinary consciousness, I am this relative personality conditioned by the environment about me and beyond. On the other side, I am that which supports this universe. There also is a sense of I ascending and descending. . . . Tentatively I applied the term "escalating self." This may be but an appearance. It does not seem to be so now, more as though both types of consciousness [were] running concurrently. But there was this kind of experience of self in the relative field, limited, restricted, conditioned by environment; the self above supporting the full universe.[89]

Mathematical characterizations of both continuity and discontinuity are probably necessary in order to understand the relationship of the relative to the transcendent.

Likening the discontinuity of consciousness to the discontinuity of a mathematical function is an example of the more general use of mathematical ideas as metaphors for understanding the transcendent. A prevalent mathematical concept used by Wolff in this way is that of infinity. For example, he has likened increases in knowledge within the relative domain to the addition of subsequent terms in a convergent series. Just as one can continue to add more terms in the series, so one can always have new ideas. However, just as it is not possible to obtain the sum of the series simply by adding more of its terms, so it is not possible to realize the transcendent states of consciousness by having more new ideas, no matter how impressive they may be.[90]

Wolff has said that the concept of infinity is particularly useful for describing the relationship between the experiences of the first fundamental realization and those of the second fundamental realization. The first fundamental realization allowed him to be conscious in a domain that was absolute with respect to the relative domain. The second fundamental realization allowed him to function in an absolute domain that transcended both the relative domain and the absolute domain of the first fundamental realization. How is this to be understood?

Similar to the previous notion concerning convergent infinite series, the relative domain can be likened to the natural numbers, which we count one, two, three, and so on. Within the relative domain we can always count further without coming to an end. However, in mathematics, one can talk about a completed infinity. One can posit the existence of a set that contains all of the natural numbers. The consciousness of the first fundamental realization was likened to the completed infinity of all of the natural numbers. But now this process can be reiterated. One can include the completed infinity in another completed infinity, of which the first infinity is simply an element. This higher-order infinity transcends the first infinity "just as completely as the latter transcends finite numbers."[91] The construction of new infinities can itself be repeated indefinitely. Thus, in the contemporary understanding of the ordering of infinities, Wolff found an analogue for the relationship between the absolute domains of his first and second fundamental realizations.

Finally, there is a much more difficult question. Is there some way in which mathematics can be used to realize transcendent states of consciousness? Wolff has strongly indicated that this is possible but has given few practical clues. He has said, for example, that "mathematic [*sic*] is that portion of ultimate truth which descended from the upper hemisphere . . . into the Adhar with minimum distortion and thus becomes the Ariadne thread by which we may ascend again most directly, most freely."[92] How is mathematics to be this direct, free ascent?

To begin with, Wolff has made a distinction between the manner of approaching the transcendent in the East versus the West. It was his purpose to describe a manner of approach that was not a translation of Eastern methods but was indigenous to the West. Using ideas suggested by F. S. C. Northrop, Wolff distinguished between an aesthetic continuum and a theoretic continuum, each of which have differentiated and indeterminate poles analogous to ordinary and transcendent modes of consciousness. The Eastern mind has been oriented to the aesthetic continuum, whose differentiated pole is the realm of ordinary perceptual experience. The aspirant of the East seeks to move toward the indeterminant pole of the aesthetic continuum. To use an analogy for this movement, rather than paying attention to images on a soap bubble, one reorients one's consciousness to the bubble itself, upon which images can play. On the other hand, the Western mind has developed in the direction of the theoretic continuum. While science and mathematics represent its differentiated pole, Wolff has said that his contribution has been the suggestion that a spiritual aspirant seek the indeterminate pole of the theoretic continuum. In other words, he has suggested using the strengths of the Western mind, found in science and mathematics, as a means of approach to the transcendent.[93]

According to Wolff, the characteristic Western modality of consciousness was set in motion by Pythagoras, who introduced the notion of a deductive proof in mathematics, supplementing earlier empirical methods. This led to the growth of a vast theoretical edifice that has required the development of potent cognitive powers. Wolff has commented on the demands that the study of mathematics placed on him.[94] Through the use of the mind, the aspirant in search of the transcendent can succeed in reaching a point "within the higher realms of dualistic consciousness," where access to the transcendent becomes possible. However, the quest cannot be consummated by further activity of the mind. Rather, according to Wolff, what is needed in order to solicit the opening of the door from the other side is "absolute humility . . . the complete sacrifice of everything that [one] is and has." Wolff has also expressed this by saying that it is not the content of scholarship that is important but the transformation from a "self-withholding" to a "self-giving attitude."[95] If one succeeds in being admitted to the transcendent state, then one finds oneself in a fluid realm where concepts are indeterminate.

Corresponding to transcriptive thoughts, there are concepts that are both determinate and indeterminate. Insofar as they are determinate, they can be used for communication. In their indeterminate depths, they reach "into the infinite." In these cases the concepts themselves are not important. Rather, they are vessels for the transcendent. It is with this in mind that one should "use writings filled with something from above." Given that mathematical axioms can arise in the course of transcriptive thought, are they themselves containers for something from above? Is the converse possible? Can the effort to develop mathematical constructs invoke something that lies above?[96]

Theoretical activity, of which mathematics is the prototype, can lead one to the door of transcendent consciousness but cannot force it open. The door can only open if everything has been sacrificed and, with great humility, one is prepared for whatever the truth may be. "The keynote, so far as the breaking through to the transcendent is concerned, is purity. . . . Purity in . . . [the] sense of completeness of self-giving. Purity which means unmixed motive, unmixed thinking." But then Wolff went on to say that "one of the greatest lessons in purity is the study of pure mathematics."[97] Could it be that mathematical endeavor of itself has the effect of developing the self-giving attitude? That the key ingredient necessary for transcendent realization is already a by-product of the process that can take one to the edge of the differentiated theoretical continuum?

There is also a sense in which engagement in pure mathematics resembles the activity of those who have become enlightened. According to Wolff, "practically all of mathematical creation has been done [for the fun of it] by the pure mathematician sitting in his ivory tower." Similarly, those who have realized the transcendent state no longer work—they play. Their actions are a "spontaneous expression of delight." But out of that play "come the greatest creations of all."[98] Can one awaken the introceptive faculty by playing instead of working? Can one become a god by imitating the actions of gods?

In this chapter, we have surveyed the motivation for seeking transcendence, meditation techniques used for self-transformation, and some of the experiences encountered in transcendent states of consciousness. But how is one to make sense of this? Is there a rationale that can be used for putting into place the fragments about transcendent experiences? Assagioli, Baker, and Wolff all had one thing in common: they were all strongly influenced by theosophical ideas.[99] Let us turn now to have a look behind the scenes, so to speak, by developing a theosophical model of reality.

CHAPTER FIVE

# Theory

## *A Theosophical Model of Reality*

> [Pontification or quotation of one's] "secret teacher's secret doctrines" . . . can be found in endless high vapor content spiritual books and teachings.
>
> —*James Fadiman*

So far, we have seen the ways in which our knowledge can be constrained by social influences in both our everyday and our scientific endeavors and characterized its liberation using the notion of authenticity. In the last chapter, we considered the possibility that answers to existential questions may become revealed in transcendent states of consciousness. All the time, however, we have stayed on this side of the line, retaining an objective and critical attitude toward the subject matter. However, what might it look like on the other side? If, like Alice, we could step through the looking glass, what would we find? In this chapter and the next, an effort is made to give more of an account from the other side. We will start with a description of reality loosely based on the theosophical tradition. Then, in the sixth chapter, some of the experiences associated with spiritual aspiration will be considered.

We can adopt a more formal approach to the subject matter of this chapter by saying that we are proposing a theory to explain the anomalous results in science and purported transcendent events that have previously been discussed. Is there a model that fits these data? What story can we tell that may make some sense of the substance of our discussions? In Chapter 3 we considered the limitations of a materialist interpretation of reality. We could persevere in seeking to fit that theory to the data, perhaps with some extensive modifications. Alternatively, we could choose a specific religious worldview, such as that of Buddhism or Christianity, and explain events in terms of the constructs of that religion. There are also less known cosmologies.[1] We could also develop our own theory, perhaps with lots of mathematics in it. However, in this case, let us use a simple model that essentially consists of a summary of some of the ideas found in the theosophical tradition.

Why pick theosophy? To begin with, it is compatible with some of the ideas already presented in this book. Assagioli, Baker, and Wolff were all influenced by theosophy. In addition, picking theosophy would be consistent with an effort to use an interpretation of reality that is at least partially suited to a Western, nonreligious approach to understanding life. It has also been an alternate model of reality that has had "important consequences" both within Eastern and Western cultures.[2] Theosophical ideas can be used to develop a reasonably articulated theory, sufficiently unlike materialism, that can serve as a second eye, allowing for some depth-perception of phenomena of interest to us.

Is it appropriate to posit a model that, as will be seen in later sections of this chapter, includes fantastical constructs that do not readily lend themselves to verification? There are two main reasons for not letting this become a deterrent. First, such models exist in both the natural sciences and the social sciences. For example, in physics there are superstring theories, which entail highly abstract constructs that cannot exhaustively be subjected to experimental verification. This has not prevented an enormous amount of effort on the part of physicists and mathematicians to develop them or to consider them as correct explanations of reality.[3] Similarly, in cognitive science, connectionist models of the mind float in a theoretical domain between the neural substrate of the brain and actual experiential events.[4] While one could argue that it is not appropriate to propose fantastical and untestable theories, that does not justify arbitrarily prohibiting the development of a theosophical model.

Second, the purpose of generating models is to provide as accurate a map of reality as possible. It may be that the fantastical constructs of theosophy turn out to have correspondences in reality when the appropriate effort is made to test them. Indeed, what appeared to be mythologies have successfully guided investigation in the past. For example, there were myths about the source of the Nile River that were sufficiently detailed to guide exploration, allowing it to be "found." For the natives who lived near the source of the Nile, it was not a myth but an actual place in their environment.[5] Similarly, those who have contributed to the theosophical literature have claimed that what sounds to us like fantastical notions are simply the elements of their lives. By considering what they have written, aimless searching can be replaced by a potentially fruitful research direction.

One other thing should be mentioned before starting this exposition. Sometimes when complex material is presented, inaccuracies are introduced in the interests of giving an account that is sufficiently broad in scope. Subsequently, one may come along and dismantle parts of the original version and replace them with alternative accounts. I have repeatedly had to resort to simplifications in presenting the material in this book and will necessarily continue to do so for the exposition of a theosophical model. Moreover, I have not avoided contradictory data or elements of a theory. If one tries to stay true to phenomena as they present themselves and if reality is greater than any logically consistent theory of it, then contradictory material is inevitable in an account such as this. The effort here is not to impose consistency but to encourage mutual exploration in order to facilitate the deepening of understanding.

### Theosophy

The term "theosophy" is derived from Greek and refers to a philosophy that espouses direct knowledge of God. This term was adopted on September 13, 1875, by a group of people interested in occult topics who would gather in the apartment of Helena Blavatsky, a Russian immigrant living in New York City. They formed the Theosophical Society to counteract the perceived inadequacies of both traditional religion and science and to perpetuate mystical traditions of the past.[6] The first official meeting was held on November 17, 1875, with Henry Olcott, a New York lawyer, as president and speaker. Within a year, the society was essentially inactive but subsequently was revived with the formation of sections in different countries as well as numerous splinter groups that charted their own course. Blavatsky has been identified with the turbulent origins of the Theosophical Society and its early ideology. In her writings, and particularly in *The Secret Doctrine*, Blavatsky purported to have accessed ancient documents and the "Masters," perfected human beings who are the custodians of the "ageless wisdom," allowing her to present a cosmology of epic proportions.

In 1915, Alice Bailey, a disillusioned Christian missionary, read *The Secret Doctrine*[7] and subsequently became a member of the American section of the Adyar faction of the Theosophical Society in which she "quickly rose to a position of influence."[8] However, "just as I thought I had found a centre of spiritual light and understanding, I discovered I had wandered into another sect."[9] While retaining her membership in the Theosophical Society, together with her husband she started her own organization in 1923, called the Arcane School, which has claimed to have had hundreds of thousands of students.[10]

Bailey claimed that in November 1919, while sitting alone on a hillside, she heard the voice of a Tibetan lama, who asked for her cooperation in writing a series of books. Initially she refused, not wishing to become involved in psychic activities, but subsequently agreed when she felt assured that her integrity would be respected. When writing, Bailey would achieve a point of "intense, focussed . . . attention," such as that developed in meditation, which would allow her to "register and write down the [carefully formulated and expressed ideas] of the Tibetan."[11] This process resulted in the writing and publication of nineteen books concerned with various spiritual matters.[12]

Baker had read Blavatsky's *The Secret Doctrine* and Bailey's *The Light of the Soul* when, in 1951, he had the spiritual experience described in the last chapter.[13] Subsequently he has produced a series of books and cassette recordings giving his ideas concerning the nature of consciousness and reality. The writings of Bailey and the writings and lectures of Baker form the basis for this chapter. However, practical problems are encountered when trying to summarize this material. Besides their sheer volume, the ideas presented are complex, often probably symbolic and sometimes ambiguous.[14] Furthermore, in Bailey's and Baker's writings, the reader is encouraged to "ponder" and "brood," so that the information conveyed by the words appears to be only the beginning of the knowledge at which it is aimed.[15] For our purposes, we can regard these cosmologies as speculative interpretations of reality. Rather than trying to faithfully reproduce them, I will describe a version of theosophy based largely on Baker and Bailey but modified by my own imagination.[16] Information that has been drawn directly from Bailey, Baker, or other sources will be indicated as such in the text or notes.

### Levels of Consciousness and Reality

From a theosophical point of view, what we normally identify as the physical world is only part of the lowest of a number of levels of reality, as illustrated in figure 4.[17]

| | | HIGHER | | | |
|---|---|---|---|---|---|
| ETHERIC | | MENTAL | | | |
| | | | | | |
| | ASTRAL | | BUDDHIC | ATMIC | MONADIC |
| | | LOWER | | | |
| PHYSICAL | | MENTAL | | | |
| | | | | | |

FIG. 4. Levels of Consciousness and Reality

Each of the columns is a plane of reality, with the outermost planes, those of greatest density, on the left and the innermost planes, those of the greatest refinement, on the right. There are more planes to the right of the ones shown here, but these are so far removed from ordinary experience that I have left them out. Within each column, the seven rows are seven subplanes, with the outermost on the bottom and the innermost at the top. All but the column on the far left could be thought of as being stacked vertically, with the lowest row of a column placed directly on top of the highest row of the column immediately to its left. Although reality is organized along a spectrum, similar interpenetrating patterns recur, allowing for the representation given in figure 4.[18]

In the lower left-hand corner are the three subplanes that correspond to the physical world as we ordinarily experience it. From the bottom up, these are the solid, liquid, and gaseous subplanes. These subplanes represent the degrees of density of physical matter. Note that this physical matter is not independently real; rather, it is the outermost level of the spectrum of the expression of being.

The other four subplanes making up the lowest level of reality are the etheric subplanes. As the atom is broken apart, science progressively encounters these, starting with the lowest of the four. These are the domain of electromagnetic and subatomic phenomena.[19] Perhaps some of the most famous predictions made by theosophists are those concerning the development and use of nuclear fission, a phenomenon that has been said to belong to the etheric plane.[20] As noted in chapter 3, events at this level of reality violate the ways in which we ordinarily think about the world. While I will continue to use the word "material" to describe the constitution of the remaining planes, the usual meanings of the word "matter" already break down for the etheric subplanes.

The material of the highest subplane of the etheric can be thought of as coming into existence as a result of the activity of material on the lowest subplane of the next plane of reality, the astral. In figure 4, the subplanes of the astral plane have been drawn with dotted lines to indicate their less discrete separation. The substance of this plane is experienced as affect and is responsive to the imagination.

The next plane, the one made up of mental matter, is again divided into two, the higher and lower mental. In figure 4, the thick line running between the astral and higher mental, between the higher mental and lower mental, and between the lower mental and buddhic planes represents the boundary between the lower three worlds on the left, which are conditioned by space and time as we know them, and the higher worlds on the right, which are characterized by eternity. These higher worlds include the higher mental, buddhic, atmic, and monadic planes. Some of the qualities of these higher planes will become apparent when I refer to them later in this chapter.

These levels of reality can also be characterized symbolically as earth, water, air, and fire. Earth refers to physical-etheric matter, water to astral, air to lower mental (and sometimes to the higher mental and/or buddhic), and fire to anything to the right of the space-time boundary (and sometimes to the mental). In this sense, everything is made up of earth, water, air, and fire.

These planes of reality are not empty. They are teeming with life that constitutes their material or utilizes it to take various forms. The first kingdom of nature is the mineral kingdom, which is made up only of physical and etheric substance. The second kingdom is the plant kingdom, which is made up of physical-etheric as well as astral material. The third kingdom, the animal kingdom, is made up of physical-etheric, astral, and lower mental matter. The fourth is the human kingdom, which is characterized by the presence of material from the lower three worlds as well as the higher mental and more refined planes of existence. The fifth kingdom is the spiritual kingdom of souls and angels and is centered in the higher planes of reality.

There are two forces at work: an involutionary one and an evolutionary one. The involutionary force moves life toward denser forms of expression, and the evolutionary force moves it toward those which are more refined. Human beings are on the evolutionary arc; their purpose is to pass out of the fourth kingdom into the fifth. Because this involves a transition from the lower three worlds to the higher, the human kingdom is a gateway kingdom. This passage becomes possible only after numerous cycles of expression on the lower planes. In other words, there is an ordered process of human reincarnation until liberation from the wheel of rebirth has been achieved.

Human constitution at the different levels of reality is illustrated in figure 5.[21]

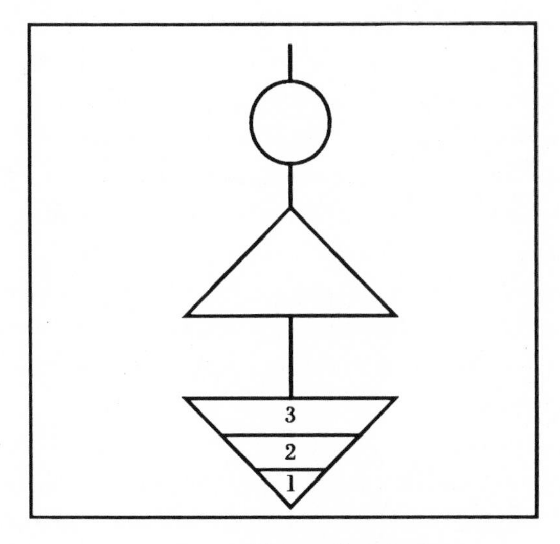


FIG. 5. Human Constitution

The personality, represented by the downward pointing triangle, functions in the lower three worlds and is made up of four vehicles of expression. The physical body is the most obvious one for most people. The etheric body is more subtle and, together with the physical, makes up the physical-etheric body, represented by the number 1. Number 2 refers to the astral body, and number 3 is the lower mental. Outer bodies are interpenetrated by inner ones, in a manner analogous to the way in which water permeates sand, so that these bodies are spatially and temporally coexistent during the normal waking state. Similarly, the spiritual triad, found on the atmic, buddhic, and higher mental planes and represented by the upward pointing triangle, interpenetrates the personality; and the spiritual triad, in turn, is interpenetrated by the monad, a cell in the body of God, which is represented by the circle. A center of focus for the attributes of the spiritual triad is located on the lowest of the three subplanes of the higher mental plane; this has sometimes been called the "soul," although I refer to it as the "transpersonal self." Its location has been indicated in figure 4 by a sun. The superconscious region in Assagioli's egg diagram corresponds to the domain of the spiritual triad, and the higher self coincides with the transpersonal self.

The line connecting the circle to the upward facing triangle represents a thread[22] that comes down from the divine on higher planes of reality to connect together all of the bodies. Thus, the monad, the spiritual triad, and the personality are nested vehicles of expression of spiritual being. However, we are generally not aware of this connection. Indeed, the point of existence is to consciously retrace the path of the thread from the bottom up to the top. On the way up, one becomes aware of the existence of the transpersonal self. Subsequently, a link, known as the "rainbow bridge," can be established between the higher mental body and the lower mental body, whereby one gains access in consciousness to the qualities of the spiritual triad.

It is interesting to speculate about Wolff's transcendent experiences in light of these schemata for the human constitution and spiritual development. His determined drive toward the root would have resulted from the influence of the transpersonal self. The cognitive and affective attributes of the first fundamental realization would correspond to the qualities of the higher mental and buddhic planes, respectively. The second fundamental realization was characterized by power—indicative of conscious penetration of the atmic plane, the plane of transpersonal will and purpose. That which is ponderable belongs to the lower three worlds, while the substantial is that which lies beyond the space-time boundary. The changes in self-identity result from crossings of the rainbow bridge by consciousness and from alternating identification with the personality and the spiritual triad. More generally, experiences of enlightenment consist in intrusions of the qualities of the spiritual triad into the personality or in the elevation of awareness to the higher planes of reality, which is made possible by the rainbow bridge.

Let us go back to the bottom and consider the etheric and astral bodies in greater detail.

**Etheric Phenomena**

The etheric vehicle interpenetrates and is slightly larger than the physical. Each body has organs appropriate to the nature of the plane on which it is found. Thus the etheric body, which is analogous to an electromagnetic matrix, has foci of energy that play a role similar to that played by physical organs in the physical body. These foci are interconnected by lines of energy that correspond to the meridians used for acupuncture in Chinese medicine.[23]

Because the etheric body is a source of form and energy for the physical body,[24] its quality of functioning has implications for physical health. Hiroshi Motoyama purportedly developed machines for measuring the characteristics of the electric field at the surface of the body and used them with five thousand subjects to correlate differences in the functioning of the subtle energy system with "the results of conventional medical examinations and [subjects'] reported symptoms."[25]

We can speculate about other practices and methods of healing whose primary focus may be the etheric body. For example, postures held for a period of time in hatha yoga[26] could create patterns of the etheric matrix, so that associated types of energy are accumulated or released. Similarly, some of the movements in the martial arts—such as those of the Chinese martial art t'ai chi and Japanese martial art aikido[27]—may work with the energy of the etheric body. The same energy could be used both to destroy and to heal.[28] The system of healing developed in nursing known as "therapeutic touch"[29] may be primarily etheric in nature. Homeopathic medicine involves the use of remedies prepared in such a way that sometimes there is not even a single molecule of the therapeutic substance left in the medication given to a patient.[30] Controversy ensued after empirical studies showed that such highly dilute substances could nonetheless evoke an appropriate response from the immune system.[31] The mechanism for these effects may also lie at the etheric level.

There are seven major centers of energy[32] in the etheric body, aligned with the spine just outside the physical body. From lowest to highest, they correspond to increasingly refined levels of consciousness. These seven centers are connected vertically by three tracts running from the base to the top of the spine, as illustrated in figure 6.[33]

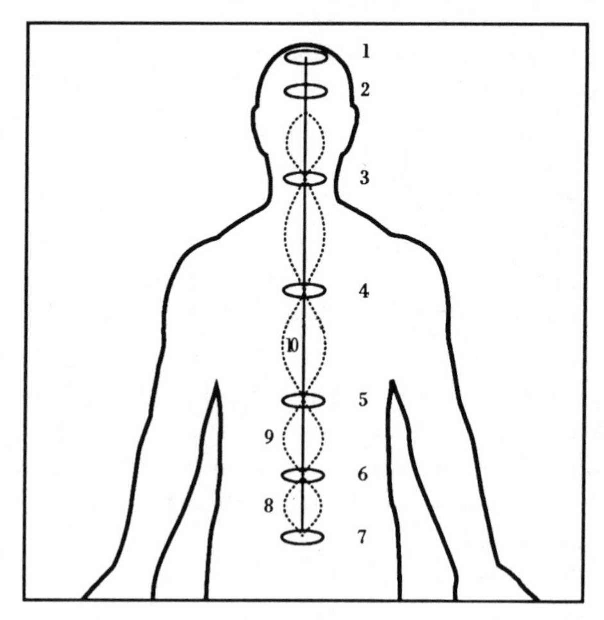


FIG. 6. Etheric Centers and Tracts

There are three centers below the diaphragm corresponding to the life of the personality in the lower three worlds. The four centers above the diaphragm are more concerned with creative and transcendent aspects of a person.

Number 7 in figure 6 designates the center at the base of the spine, which intimately connects human beings to the rest of the natural world. Instincts operate at this level, allowing birds to find their way south for the winter and native people living in tribal societies to have immediate awareness of their cohorts. It is here that the latent energy of matter—known as kundalini—resides, which, it is recommended, one not seek to arouse until one can do so "by the united authority of the soul and personality, integrated and alive . . . in full consciousness."[34] In other words, kundalini is supposed to be aroused after one knows what one is doing, not in order to find out what is going on.

The sacral center, number 6, is located at the lower part of the back and is concerned with sexuality and "form-building" more generally, thus linking it to the physical plane of reality. It is through the solar plexus center, number 5, that desires and emotions are mediated and the astral body finds an "[outlet] into the outer world."[35] Since desire functions as a strong conditioning factor for many people, this center is important in the life of the personality for much of humanity.

The energies coursing through the centers are amenable to transmutation. In particular, the energies of desire, associated with the solar plexus center, can be expressed as love through the heart center, number 4, which is "located between the shoulder blades." Thus a distinction can be made between a love that is self-serving and a love that seeks the well-being of the greater community. The altruistic love of the heart center finds practical expression through good will, a sense of responsibility, and service to humanity.[36]

The throat center, number 3 in figure 6, is found at the back of the neck and mediates thinking on the lower mental plane and is well developed in many people. It is the recipient of sublimated energies from the sacral center, so that procreation is replaced with intellectual creativity.[37] There is also a connection between breathing and thinking, which is suggested by the location of the throat center. This was noted at the turn of the twentieth century by William James, who maintained that there was no such thing as consciousness, that what one does have are thoughts, but that thoughts are nothing other than the breath moving between the glottis and the nostrils.[38] This conclusion would have been substantiated not only by James's own introspective activity, whereby he noted that the only movement he could find while thinking was the movement of his breath, but also by his knowledge of Eastern religions, in which breathing has been associated with alterations of consciousness.[39]

While the throat center is related to the personality, there is a higher throat center, also called the "alta major" center, that is more concerned with spiritual expression. Its dual location is best visualized by imagining yogis who wear large circular earrings in order to facilitate the movement of its energy. The location of the two earrings corresponds to the two wheels of the higher throat center. As this center develops, however, it is manifest as a single winged wheel lying halfway between the two outer wheels. The relationship between the lower and higher throat centers in the etheric body is analogous to the rainbow bridge between the lower and higher mind. While the lower throat center governs cellular energy consumption, the higher throat center regulates prana, which is energy stemming from the first etheric subplane. The greater the expression of prana through creative service to humanity, the greater the infusion of energy. A fully functioning higher throat center reveals its activity in a person as enthusiasm.[40]

The brow center, number 2 in figure 6, is between the eyebrows, in the plane of the face, and eventually becomes the center for the functioning of the integrated personality. It can receive energy from the heart center. Energy from the center at the base of the spine can be transmuted to the head center, number 1, which is in the plane of the top of the head. It "registers purpose" and distributes energy characterized by will.[41]

The centers can be not only active or inactive. They can also be upright or inverted; they can turn clockwise, counterclockwise, or simultaneously in both directions; and they can function in a coordinated manner. For example, the winged wheel, brow center, and head center spin in mutually perpendicular planes. When their functioning becomes coordinated, they define a single, integrated sphere.[42]

In the long course of spiritual development, lower centers are progressively neutralized, and higher ones are brought into activity. This has implications for what constitutes evolution and regression. For example, the center at the base of the spine and the centers above the diaphragm provide direct connections to the environment, but through entirely different mechanisms. Polarization at the solar plexus center, which is the situation with much of humanity, creates the greatest degree of isolation.[43] In order to reconnect and experience extrasensory perception, one could reactivate the lower centers. These are appropriately used by shamans in tribal communities in order to serve their cohorts by providing information or healing that would otherwise not be possible.[44] This model predicts, however, that while the use of shamanistic trance-inducing rituals by those in an industrialized society could lead to strange experiences resulting from the reactivation of lower centers, this reactivation may serve to retard rather than accelerate the approach toward transcendent states of consciousness. This is not a recent idea. The Chinese Taoist Chuang Tzu, writing in the fourth century B.C., made the distinction between the shaman and the "Holy Man," with the "Holy Man" able to understand the spiritual dimensions of the world that the shaman overlooks.[45]

There are three etheric tracts connecting the centers as illustrated in figure 6. Pingala and ida, numbers 8 and 9, respectively, form two helixes around sushumna, number 10, which runs straight up the middle, giving the appearance of the caduceus made up of two snakes entwined about a staff.[46] As energy courses past the centers, it sets them in motion. Pingala has been associated with matter, ida with consciousness, and sushumna with spirit.[47] There are also correspondences between pingala and suffering and between ida and pleasure, so that sushumna represents release from both suffering and pleasure. There are numerous impediments to the free flow of energy in these three tracts, which are dissipated as a result of "the livingness of the individual centres."[48] Forceful warnings in the theosophical literature caution against attempts to tamper with the energy system or to raise the kundalini for the purposes of satisfying one's desire for spiritual achievement.[49]

### Astral Phenomena

Astral material is such that it can easily be shaped by the human imagination.[50] Because the nature of matter on the astral plane is to "form globules,"[51] whole spheres can exist furnished with environments corresponding to specific themes. For example, the characters and events in a novel are brought to life in the astral plane by their author. Then every time someone reads the book, these thought-forms are strengthened. In Lewis Carroll's *Alice's Adventures in Wonderland,* for example, Alice is a thought-form on the astral plane. As someone thinks about Alice, her thoughts encounter the astral Alice and increase Alice's vitality as a thought-form. As people stop thinking about something, the thought-form associated with it on the astral plane atrophies and eventually disintegrates.

According to this model, schemata and scripts are not isolated within each individual but exist in the atmosphere of the astral plane, where they are accessible to everyone. Not only are we conditioned as children through our relationships with parents; at some point we click into and enact normative thought-forms. The unwitting utilization of shared exemplars on the astral plane is the mechanism underlying inauthenticity. Not the least of these exemplars are those belonging to religious traditions. Because of the extent of human preoccupation with religious icons, they are continuously re-energized. That is not to say, for example, that there is no Buddha or Jesus, just that the images of Buddha or Jesus that one encounters in one's imagination are unlikely to be anything other than astral thought-forms, shaped by human beliefs and desires.

Scientists are not exempt from this process. In particle physics, theorists postulate the existence of subatomic particles with specific properties that would be necessary for the consistency of their mathematical theories. In terms of this model, such particles could be created as thought-forms on the astral plane. Under some conditions, that which is created as a thought-form on the astral plane can descend into the physical-etheric plane.[52] Is it possible that the increasingly greater number of particles being found by experimental physicists could simply be the result of their capacity to materialize them? In that case, the number and types of particles that are found will be limited only by the limitations of the imagination and collective beliefs of physicists. While this is highly speculative, it does illustrate the degree to which this model potentially conflicts with our usual interpretations of reality.

The lower mental, astral, etheric, and physical bodies are sheaths formed around the transpersonal self. When a person dies, she successively divests herself of these vehicles. Thus at the time of physical death, she leaves the physical and etheric bodies and finds herself in the astral worlds. In addition to the bubble-worlds consisting of fantasized characters and venues, there are astral worlds corresponding to the physical world. According to the theosophical theory, it is here that most people find themselves after death, often without realizing, for a while, that they have in fact died.

There are three dimensions of experience that one can consider when analyzing any altered state of consciousness, including death: one's ability to cognitively identify what is happening, the degree of control that one can exercise, and the intensity of one's sense of being.[53] When someone has died, even though she may realize that something has happened, she may not be able to identify it as death. To the extent that the critical faculty necessary to make that determination depended upon the physical brain, it is no longer available. However, if the mind and transpersonal self have been developed to some degree, they can offset the loss of the physical brain.

There is also loss of control, in that one finds oneself in a compartment of the astral that coincides with one's disposition. During her lifetime, a person would have had specific desires, fears, and moods, which constituted the astral body. When she died, she would have been drawn to an element in the astral world corresponding to the coarsest of these emotions.[54] Thus someone who cultivated deceit during her lifetime would end up with others who had had a similar disposition. After some duration of experience, this coarse material would slough off, and she would rise to a more refined level of the astral plane. At some point, a person may end up in heaven worlds whose ambiance coincides with her expectations of heaven. However, this, too, is only a temporary state; eventually, a second death occurs, whereby a person dies to the astral plane.

The more one has functioned deliberately as a conscious being during one's lifetime, the more likely one is to have a greater degree of control over one's astral state. Those who have largely mastered life in the lower three worlds can range at will and may choose to function in the astral plane in order to accomplish specific goals. Instrumental transcommunication, discussed in chapter 3, can be explained in terms of this model. Those who have died, together with nonhuman beings, have set up a transmitting station in an astral world to communicate with those who are on the physical plane.

Those who believe that life does not end with death sometimes think that the deceased are accoutered with all kinds of attributes at the moment of death. Death, of itself, does not bring about personality changes. Rather, the qualities that a person cultivated during physical incarnation are carried over at death. One does not, for example, become benevolent or wise just by dying. In addition, the intensity of one's sense of being—that is, the degree to which one experiences oneself and events as real[55]—after death will probably be consistent with what it was before death. However, there are events that occur during the process of dying that may be accompanied by a heightened sense of reality. Some indication of these may be gathered by comparing elements of this theosophical model with some of the features of near-death experiences.

Some aspects of theosophy and near-death experiences are in agreement. First, the abstraction at the time of death from the physical and etheric bodies would correspond to a person's sense that she had left behind her physical body. Second, when consciousness is shifted from the physical-etheric level to the astral, an aperture may sometimes be seen.[56] This would correspond to the journey through a tunnel described in near-death experiences.[57] Third, Bailey has maintained that the relinquishing of the physical and etheric vehicles draws the attention of the transpersonal self to the personality. One of the effects of this may be a sense of timelessness during which the events of the past life are displayed.[58] This would coincide with the life review that sometimes occurs during a near-death experience.[59] Fourth, contact with the transpersonal self can result in an effusion of light.[60] Correspondingly, a loving light is sometimes perceived during a near-death experience.[61] These are some of the ways in which a theosophical theory fits the phenomena of near-death experiences.

There are, however, some problems in modeling near-death experiences using theosophy. For example, some who have had near-death experiences have reported that they have seen their own bodies.[62] How does this seeing occur? It is one thing to say that a person can see the various articles and people in a room while she is outside of her body; it could be argued that, using etheric or astral eyes, she is seeing the etheric or astral counterparts of these articles and people. However, that argument does not work for a person's own body because the body with which she is looking has been abstracted from the physical sheath. A person looking down at the resuscitation process during a near-death experience should just see real etheric or astral people working on an invisible body.[63]

In this theosophical theory, one does not have to be dead or nearly dead in order to travel in the astral worlds. During sleep, attention shifts from the physical to the astral plane. Dreams are just sequences of events played out in astral matter. Lucid dreams are dreams in which one realizes that one is dreaming. They are usually accompanied by increased control of dream contents and by a heightened sense of reality.[64] While one is awake, the location of one's higher vehicles normally coincides with that of the physical body. In sleep, however, the etheric and astral vehicles go out of alignment with the physical, and the astral body can travel in the astral worlds. Out-of-body experiences, as these astral travels have been called, have been discussed both academically as well as more generally.[65] From a traditionally scientific point of view, out-of-body experiences have been said to result from the separation of one's model of the world from the sensory input necessary to ground it, so that one ends up imagining oneself outside of one's body in alternate models confabulated from schemata stored in memory.[66]

Psychism is another way of gaining access to astral material. While this may include the use of divinatory devices or procedures—such as tarot and automatic writing[67]—at the core is an entering into a passive state whereby a person renders herself susceptible to astral influences. From a theosophical point of view, psychism can jeopardize the integrity of one's vehicles of expression and hence is considered highly undesirable. The model predicts not only that one is unlikely to find that which is real amid the imaginary fragments but also that one opens oneself to the possibility of possession by astral entities. The key lies in the attitude with which astral material is approached. The integrity of the vehicles of sensation and expression needs to be retained. Some of the problems that could arise from the cultivation of a more passive attitude will be mentioned in the next chapter.

### Initiation

The purpose of the theosophical literature has been primarily to assist those who are trying to be spiritual. Much of the focus has been on the subtler vehicles and how their qualities become manifest in the course of one's spiritual unfolding. In this section, I will present some of the phenomena associated with spiritual aspiration as conceptualized within a theosophical framework; experiences encountered in the process of self-transformation will be discussed in the next chapter.

To begin with, an aspirant is encouraged to reorient herself from an emotional to a mental polarization. That is, rather than acting with the center of her consciousness at the emotional level, she is to relocate that center to the lower mental level. This is contrary to many people's ideas concerning spirituality. According to them, the whole problem is that we think too much, and that spiritual effort lies in recovering suppressed emotional aspects of ourselves. From the point of view of this theosophical model, however, there is nothing spiritual about emotions in and of themselves. And while there is nothing spiritual about the lower mind, either, it has the advantage of forming one end of the bridge between the personality and the higher triad.

There is a symbolic way of understanding another of the benefits of mental polarization. Each etheric center has associated with it a number of animals that symbolize its characteristics. One of the animal symbols of the throat center, which mediates the activity of the lower mind, is the camel. A camel, apparently, is obstinate and hard to train. Once domesticated, however, its strengths can be exploited. Because of its capacity to store water, a camel can plod for long distances in the desert. For a long time, access to the transpersonal self occurs intermittently, marked by intervening periods of aridity. Contact with the transpersonal self is symbolized by drinking at oases, and the intervening aridity, by the desert that one traverses. The emotionally polarized person would alternate wildly between elation and despair, depending upon whether she were in the presence of spirit or were cut off from it. The mentally polarized person can recognize such a pattern and deliberately seek to modulate her emotional reactions, so that the sense of well-being arising during times of inspiration does not get used up all at once but spreads out into periods of effort when self-discipline is all that keeps her going. Like a camel, the mentally polarized person can plod from one oasis to the next.

There is a sense in which emotions can be spiritualized. An affinity exists between the astral and buddhic planes, so that knowledge at the buddhic level can be expressed as intuition in the astral body. However, strong fears and desires in the astral vehicle distort the feelings that are prompted from the buddhic plane, and the resulting intuition may be so perverted as to bear virtually no resemblance to the original spiritual impulse. In addition, fears and desires show up in the guise of insistent promptings. The greater the clarity of the astral body, the greater the likelihood that one's intuitions have spiritual origins and that they bear some relationship to the truth. At some point, one may be able to rely on one's feelings. However, one generally has to struggle to achieve this; it does not occur simply because one believes that it does.

Different metaphors can be used to characterize the process of self-transformation. The goat can be used for depicting the aspirant in the beginning stages of spiritual effort. Initially, one is a sheep among other sheep in a flock, symbolizing an inauthentic mode of being. A sheep, presumably, does not make her own choices but does whatever the others do. At some point, however, there is a change. One develops the courage to think for oneself. Then one is no longer a sheep but a goat. And rather than wandering with the flock, one purposively sets off in one's own direction. Whereas sheep have to be rescued if they fall into a crevice, a goat can scramble up the side of a rocky cliff. As mountain peaks and the north symbolize the spiritual,[68] scaling the cliff corresponds to the exacting task of seeking out the transcendent. The goat, leaving the flock and scrambling up the cliff, is thus a metaphor for spiritual aspiration.

What happens at the mountaintop? From a theosophical point of view, there are five "expansions of consciousness,"[69] which are referred to as initiations. These are planetary in scope and independent of a person's field of endeavor and religious orientation. The first and sometimes the second initiation can occur without the person's identification of the event as such. It is also possible that one or both of these first two initiations has been taken in a previous lifetime and not recalled in the present life.

It could be said that the first initiation is the culmination of a religious approach and the beginning of a scientific one. Mystical tendencies are replaced with the need to know "the workings of that great Intelligence which created the manifested universe."[70] A sense of dualism is brought about by the knowledge that there is something else other than just the personality; the purpose now is to find this other. There is a deep desire to search for the truth. Attachments and old beliefs and ways of thinking are swept aside. The initiate may have difficulty keeping her feet on the ground. There is an interest in esoteric teachings that purport to get underneath the surface of things. The period after the first initiation is characterized by a process of reorientation to a life of spiritual aspiration, and the possibility of turning back disappears.[71]

A number of phenomena may occur within the scope of the initiate's experience. There may be a period of increased light within herself or in the world about her. She truly loves others. Both physiological and emotional sensitivity is increased. Making decisions becomes more difficult with recurrent interference from the initiate's conscience. She seeks to discipline herself in order to live a life of greater beauty. Karmic retribution, a cosmic process of adjustment for misdeeds, is speeded up, so that within hours of a blunder, the initiate is confronted with the consequences of her actions.[72]

Not only has the initiate been initiated into a new pattern of living; she creatively initiates groundbreaking activities within the sphere of her service to humanity. She is not interested in maintaining old systems but tries to introduce new ones and is not deterred if they are initially less successful than the old ones. In place of static teachings, she presents teachings that are heretical from the point of view of the culture in which she finds herself.[73] Given the initiate's reorientation to a search for truth, her increased sensitivity, her effort toward self-discipline, and her use of new methods within a social milieu that abhors change, it is not hard to see that her life would be difficult at this stage.

These characteristics, restated from obscure sources, may seem like the epitome of mystical fantasizing. There are those, however, for whom they strike a chord. *Yes, that is what it's like. That's exactly what happened to me.* And a person thus has a map on which she can point and say, *this is where I am.* The map is not the terrain, and the map may not accurately correspond to the terrain. But for some people, it is a relief to find a model that includes at least some of the phenomena that they have experienced.

While the first initiation is concerned with the birth of a new person—symbolized by the element earth and related to the sacral center—the aim of the second initiation is release from emotional control, which is symbolized by the dissipation of fog and correlated with the ascent of energy out of the solar plexus into the heart center. Symbolically speaking, the dedication to spiritual effort that characterizes the first initiation draws forth spiritual fire, which strikes the water of one's emotions, creating a mist. The emotional upheavals that constitute this mist are dissipated by devotion—initially to an individual or an idealistic project but ultimately to the treading of the spiritual path and "unswerving attachment to service."[74]

Someone who has experienced the second initiation would have the capacity to withstand tremendous criticism and stress. Her life would be exemplified by humility, the recognition of divinity within each person, intellectual capability, and dedication to a life of service. Such a person would have experienced not just transient periods of enlightenment but a deeper illumination, which is reflected in the quality of her teaching.[75]

While the time between the first and second initiations is a time of difficulty, that between the second and third initiations would be "a period of intense suffering."[76] The aspirant would work in the dark, guided by her mind with occasional inspiration from the transpersonal self. This would culminate in the third initiation, at which time, symbolically, she would move *"out of the fire, into the cold,"*[77] and the transpersonal self would become the dominant factor in her life. The brow center would be the focus of attention at the time of the third initiation, and it would be used to redirect the spiritual energies that would pour through her after that initiation.[78]

### The New Age

There is one more story in the theosophical repertoire that is relevant to our basic discussion of consciousness and reality. It is concerned with long-term events that shape human history and includes the notion that at this time, humanity is stepping out of one age into a new one.

In order to properly discuss the idea of a new age, we need to make a small detour into astrology. The term "astrology" refers to various systems of relating celestial bodies to human affairs. One rationale is that the etheric counterparts of stars, planets, and moons affect the etheric body of the planet earth and its inhabitants, thereby creating a correspondence between astronomical and planetary events. Within the scientific community, astrology is considered to be nonsense. Research that has been conducted in order to determine the validity of astrology has largely served to create greater controversy.[79] Fortunately, the usefulness of the notion of a new age does not rely on the existence of actual celestial influences, since even giving a description of what is meant by the new age already exposes problems with contemporary orthodox astrology.

From the point of view of the earth, the sun appears to move against a background of fixed stars. The path it traces in the sky in the course of one year is called the ecliptic. Within eight degrees on either side of the ecliptic are twelve constellations, which are defined by specific fixed stars. The ecliptic can be divided into twelve equal arcs of thirty degrees, each of which is associated with one of the constellations, giving rise to the twelve signs of the zodiac. The earth is tipped 23½ degrees relative to the ecliptic, so that the celestial equator, the projection of the earth's equator onto the background of fixed stars, intersects the ecliptic at two points—the vernal and autumnal equinoxes. Due to a number of factors, including the gravitational attraction of the moon and sun on the equatorial bulge created by the rotation of the earth, the rotational axis of the earth wobbles relative to the fixed stars. Because of this precession, the equinoxes are slowly moving backward through one sign of the zodiac every 2,148 years.[80] In this context, an age is the time that the vernal equinox spends in one of the signs of the zodiac. For the past two thousand years or so, the vernal equinox has been moving through the sign of Pisces and sometime around now is entering the sign of Aquarius. The term "new age" thus refers to the imminent age of Aquarius.

In the above paragraph, I have identified the signs of the zodiac with the actual constellations in order to give meaning to the notion that we are passing out of the age of Pisces into the age of Aquarius. The movement of the vernal equinox through the constellation Pisces and toward the constellation Aquarius is an actual astronomical event. However, this does not coincide with popular astrological notions. The vernal equinox, when the sun is over the equator, occurs at about the same time each year, in March. Hence, someone born in late March would be born when the sun was against the background of Pisces or perhaps Aquarius. In another two thousand years or so, a person born in late March would be born with the sun against the background of the constellation Capricorn. But in orthodox Western astrology, these people would still usually be identified with the zodiacal sign of Aries. What is going on?

Astrologers have made a correction for the movement of the vernal equinox each year, so that the ephemeris, or table of planetary positions, that they use for astrological calculations would list the positions of the planets relative to a zodiac that always starts with Aries at the vernal equinox.[81] In other words, the zodiac is defined by splitting the ecliptic into twelve equal portions and labeling them Aries, Taurus, and so on in an eastward direction starting at the vernal equinox.[82] Thus the signs of the zodiac have been disconnected from the constellations, which presumably gave them their meaning. Hence one ends up with the situation, for example, where one is born with the sun in Sagittarius according to orthodox Western astrology, while the sun is against the background of the constellation Scorpio in reality. This discrepancy raises questions about the possibility of an astronomical basis for astrology.

However, the discrepancy in astrology does not affect the notion of ages as we initially presented it. If the astronomical rationale for ages appears thin, then the astronomical terms can simply be viewed as metaphorical expressions of certain ideas that may have some correspondence to reality. In other words, we can regard astrological notions as constructs that may be useful for describing elements of human activity and experience.

According to this story, during each age certain ways of being spiritual are more propitious than others. Whenever ages change, advisable strategies for spiritual aspiration also change. However, being used to the previous methods, people may be reluctant to accept the new ones. For example, the age of Taurus ended with the beginning of the age of Aries. It was presumably around this time that Moses came down from the hill with some rules of behavior consistent with a spiritual life. Symbolically, Moses' followers were worshiping a bull and did not want these laws. Law, however, was a keynote for the age of Aries and indicated a particular line of approach to the transcendent.

When the age of Aries ended and that of Pisces began, it was Jesus' turn to present a new spiritual covenant. Now it was the lawyers who implored him not to throw the system out altogether. But Jesus said that an entirely new system—one not based on rules but on the quality of people's interactions with one another—had to be put into place.

Pisces, the Fishes, is a water sign, associated with the solar plexus center and the emotions. The keynote of Pisces is belief. In order to be spiritual, one believes the right things, whatever they may be, and cultivates a devotional attitude to an object of worship. The active ingredients in accessing the transcendent in this age are the refinement and intensification of emotions. Symbolically, this would correspond to the ascension of consciousness through the astral subplanes until one was adjacent to the subplanes of higher mind and could penetrate the higher worlds laterally, so to speak. Moments of emotional ecstasy would be times of transcendent experience.[83]

However, the success of this method depends upon being able to keep the lower mind out of the picture. When using this emotional approach to the transcendent, the purpose of thinking is not to genuinely analyze or deduce some unexpected consequence of a train of reasoning but simply to rationalize what one already believes. The pastor or minister in a Christian church can start with any question and bring in the most disparate ideas, but the conclusion will always be the same. Thinking serves one's beliefs. In fact, from a devotional point of view, thinking is spiritually impotent. It is faith that has the power to produce spiritual effects.

The keynote of Aquarius is knowledge. Aquarius rules thinking, communication, and science. The lower mind, its etheric counterpart, and the physical brain are all stimulated. In this age, the tendency toward belief is being replaced with the need to know. In addition, numerous religious systems that were previously inaccessible are being made available to the public. Suddenly, it is no longer clear which religion's doctrines, if any, are correct and which of them should be believed. The effort to understand replaces faith.

But old tendencies die hard. Much of science itself operates as a religion, one in which a person is expected to believe the accepted materialistic theories for fear of persecution. Within religions themselves, the battering of traditional beliefs by ideas contrary to established doctrines leads to an exaggeration of Piscean tendencies. In place of the natural isolation from discrepant cosmologies that existed in the past, a devotee may deliberately try to segregate herself so as not to be exposed to unwanted ideas. She may insist more fervently on the legitimacy of the authority that she has accepted as the basis of her beliefs. She may become more rigid about what is the truth and what is not, lest small changes away from preferred interpretations of sacred writings unravel the entire fabric of her beliefs. The activity of the mind may be denigrated and anesthetized so that it cannot interfere with the process of producing a sealed column within which she can refine and intensify her emotions. Her emotional displays may become more desperate. But the thrust of the world is such that it continuously tries to pull her away from her objects of devotion. The final solution is to become a member of a cult, where preferred beliefs can be safeguarded against the intrusion of reason.

Problems encountered during this time of transition are aggravated by the fact that there is no sanctioned avenue for spiritual expression consistent with Aquarian needs. The Piscean methods of Western religions often fail to satisfy the inquisitive aspirant. Eastern philosophies are, to some extent, more palatable, but they often emphasize Piscean obedience to a guru. Many people turn to a new-age cocktail of religious and superstitious elements that has not been sanctioned within the greater culture. As a society, we have failed to provide an infrastructure that allows people to develop their understanding of the ultimate nature of reality, so it is not surprising that those motivated by existential questions sometimes end up turning to fundamentalism, Eastern gurus, and the new age. At least these people are trying to resolve the difficult issues with which they are faced rather than avoiding them. The real perversion lies in our society, which generally suppresses the exploration of one's own consciousness and denigrates spiritual aspiration.

CHAPTER SIX

# Self-transformation

## *Adventures and Misadventures of a Spiritual Aspirant*

> Please pray for me, Imants, not that I do the right thing, but that whatever I do, I do it with style.
>
> —*From a letter from a friend*

Let us go back to the beginning. We are merrily living our lives inauthentically, fulfilling, as best we can, the expectations of the social environment in which we find ourselves. Then something happens. Perhaps we get cancer or give birth to an autistic child, or we cannot extricate ourselves from continuous poverty. The actual events precipitating change can range widely in terms of their objective severity. It is the degree to which they are experienced traumatically that seems to make a difference.[1] In light of our crisis, it may become difficult to go through the motions of life. *What is the point? What difference does it make? Why should I keep going?* And we may start to seriously try to understand what life is all about.

This is an oversimplification of actual events. One could have been religious to begin with and develop an authentic religiosity in the process of deepening one's spiritual understanding.[2] Or one could have been religious to begin with and reject religiosity for an agnostic attitude. There may be a series of crises or none at all. Dissatisfaction may not be conceptualized in existential or cognitive terms. One could, for example, desire liberation from the limitations of one's life. However, this oversimplification is one way in which the motivation to deliberately move toward the root can be conceptualized. The question now is, from a practical point of view, What are some of the actual experiences that could occur to someone who is trying to deepen her understanding of reality? The information provided in the earlier chapters provides the framework for looking at some of the specific adventures and misadventures that one may encounter in the effort to resolve existential issues.

In writing about these experiences, I will draw on observations that I have made as a participant in various "spiritual" activities. I am aware that erroneous conclusions can be drawn from unsystematically obtained data by a single investigator. Because of my life situation and personality characteristics and the choices that I have made, I have been exposed to certain phenomena and not to others. Any conclusions that I draw would reflect my own experiences. There may be other events to which I have not been exposed that may contradict my conclusions. The alternative would be not to share the conclusions that I have drawn for myself. It seems to me that until systematic studies are carried out to the extent that they are possible, unsystematic participant observation such as mine is preferable to speculation by those who have not observed the relevant phenomena.

### Perseverance

When events occur that cause one to reexamine one's existential presuppositions, they are likely to affect social adjustment. If life seems pointless, how much effort will a person put into the actions necessary to make it through the day? Examples of changes in levels of social adjustment are illustrated in figure 7.

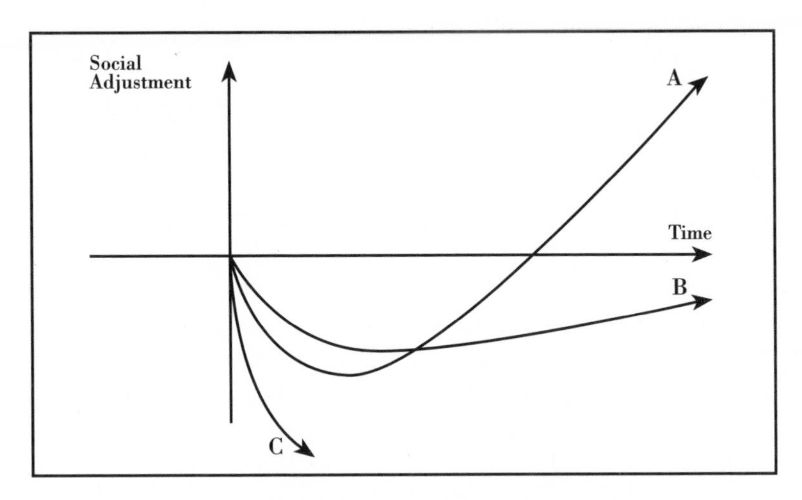


FIG. 7. Social Adjustment Following Existential Crisis

In this figure, the horizontal axis represents time, and the vertical axis indicates social adjustment. The horizontal axis has been placed at the level of social adjustment that represents normal inauthentic functioning.[3] Movement below the horizontal axis is indicative of maladjustment; movement above represents better than normal functioning. The vertical axis intersects the time line at the point of occurrence of an event that precipitates an existential crisis or the beginning of a process of deepening.

Some examples are given in this figure. The curve *C* represents someone who has difficulty functioning after a crisis and never recovers. She may suffer from depression or psychosis. In some cases, the person's psychical structure becomes fragmented. In other cases, a person may divest herself of her masks and live in a manner that is true to herself. However, her interface with other people may be so eccentric or destructive that she cannot function within society.

Unless a person is situated in a tolerant environment with ample time and resources, an existential crisis will have some impact on social adjustment. Time is needed for contemplating existence and getting to know herself better. She no longer has to accommodate only the external world but the internal world as well. Initially, there may be points of conflict between the outer and the inner. For example, the surfacing of a moral conscience may make it difficult to deceive others on a day-to-day basis if that is required in a person's occupation. Gradually, however, she may adjust her life circumstances so that they are compatible with her inner needs. Now she may again function at a normal level of social adjustment, but having taken into account her inner life, having accepted responsibility for her actions, and possibly having acquired some ability to draw on superconscious faculties. An example of this is given by the curve *B*.

Freeing oneself from inauthenticity can also free one from mediocrity. A reexamination of one's life, while it may initially lead to social maladjustment, can also eventually result in exceptional well-being, performance, and social relationships. This possibility is represented by the curve *A*.

Sometimes people imagine that spirituality is like floating down a tropical river, sipping a cold piña colada, while palm trees gently wave in the breeze. There are organizations that promise this type of tranquillity. If you chant a specific mantra or visualize specific internal processes or believe the right things, then you will easily achieve cosmic consciousness or liberation or be saved. I think that this is misleading. On the basis of my observations, for quite some time following the initiation of spiritual effort, one's difficulties become aggravated.

Why do I need to say this? Is it not obvious that life includes times of difficulty? This question brings up a number of issues. Suppose that a person with Piscean tendencies dogmatically asserts that reality is the way that she has come to believe that it is. For example, someone who has developed a career as a molecular biologist may insist that the world is essentially material in nature. Then something happens. Suppose that she has a near-death experience and comes to believe that life continues after death. *Materialism turns out to be wrong. Transcendentalism, to be right.* But what happens sometimes is that everything that was previously rejected as false now gets accepted as true. Not only does life continue after death, but there are no mental disorders, psychic surgery works, Atlanteans built the pyramid at Ghiza, Martians put the face on Mars, and Jesus wrote *A Course in Miracles.* This is not to say that there are no Martians. The point is that there has been a change in content but no change in inauthenticity. One has simply replaced one set of beliefs with another, without examining the psychosocial structures that sustain both of them. Believing in materialism and believing in everything strange are just two sides of the same coin. In both cases one may retain certainty by believing rather than examining the evidence. Thus when a smiling, well-dressed devotee presents an effortless method for reaching cosmic consciousness, this, too, may be believed.

A second issue raised by our questioning the need for a discussion of the difficulties inherent in spiritual aspiration brings up the matter of perseverance.

Suppose now that inauthenticity is gradually recognized, so that eventually everything gets called into question. One realizes that the world need not be the way that one has always believed it to be. It can be radically different. *Barušs says that everyone who tries to be spiritual experiences periods of difficulty, but what does he know? I can choose. I'm not a victim of the past. I can follow my vision and chart my own course into the future. And my vision does not include difficulties. It doesn't matter if no one else has ever floated down a tropical river sipping piña coladas. This does not mean that I can't do it. I can do what I want.*

While this attitude of self-determination is perhaps healthier than the more phlegmatic attitude of universal acceptance, there is a problem with it, which lies in its success. Through desire and the skillful use of the will, that which we visualize can, to some degree, be brought about within the scope of our experience. The mechanism whereby this occurs may not be confined to traditional methods but may include extraordinary means. For example, as discussed in chapter 3, from the results of studies conducted at the Princeton Engineering Anomalies Research Laboratory, it would appear that some physical events correspond to stated intentions without there being any known mechanism of interaction. More generally, in psychosynthesis there has been the notion that *"all inner images have motor power"* and, in theosophy, that *"energy follows thought,"*[4] so that manipulating the physical world using one's mind could lead to the realization of one's vision. Indeed, the purpose of this book is to point out the value of such self-directedness. But there is a problem with it if one tries to shape one's life without leaving enough room for the unknown. To see this, let us consider what can happen in the process of psychotherapy.

While psychotherapeutic procedures vary widely, it is possible to identify a generic pattern of counseling, which can be used for individual clients who are not overly dysfunctional. Initially, the psychotherapist provides a supportive environment for the client so that she comes to trust both the psychotherapist and her own feelings and thoughts.[5] Once such an environment has been established and the psychotherapist starts to recognize elements of the client's life that are dysfunctional, the psychotherapist will start to direct the client toward an examination of those aspects. At this point, the client may feel betrayed and resist the process. It is more comfortable to be emotionally stroked than it is to have to face oneself. The psychotherapist tries to keep enough pressure on the client to get her to work on her "stuff" without frightening her away.

Most contemporary theories of human motivation are based on the principle that we seek pleasurable experiences and try to avoid painful ones.[6] Thus we would naturally be attracted to spiritual organizations that promise pleasurable experiences with the least expenditure of effort. Through a combination of desire and volition, we bring about, at least in part, the experiences that we anticipate. At that point, we would be in a situation like that of the client who is enjoying the supportive environment produced by the psychotherapist. However, it seems to me that one of the functions of a spiritual teacher has traditionally been the same as that of a psychotherapist, namely, to direct the aspirant toward that which she is naturally inclined to avoid. Thus both psychotherapists and spiritual teachers have contradictory roles, being sources of comfort as well as of frustration for clients and aspirants.

The consumer orientation of Western culture has carried over to religious practice,[7] and, from my experience, to spiritual traditions more generally. If we do not like the product that we have been consuming, we shop around for something else in the spiritual marketplace. Given the possibility of spoiled goods, this is probably not a bad idea. Given the increasing focus on ratiocination and the availability of alternative spiritual systems that are consistent with the notion of the dawning of the age of Aquarius, this is probably inevitable. The consequence, however, may be that one only accepts those messages from spiritual teachers that fit with one's desire for well-being. In effect, then, one misses the potential benefits that a teacher can provide.

As someone who has drifted from one spiritual practice to another, dissatisfied with all of them, I can empathize with those who find themselves without instruction in their spiritual endeavors. In that case, one needs to develop discipline, patience, and perseverance in order to neutralize the natural tendencies of the personality, so that one can seek to penetrate more deeply into one's own consciousness. In particular, if one is to move past the acquisition of comfort, one has to offset, to some extent, one's natural tendency to avoid pain and seek pleasure. From the point of view of existentialism, both pain and pleasure can become relativized in the face of one's death, so that one is freed to act authentically. Using the theosophical model, one could say that the duality represented by the etheric tracts ida and pingala needs to be balanced. In Buddhist terms, an aspirant needs to make a commitment to stay with her spiritual practice and to face her "boredom, impatience, and fears."[8] In order to proceed on her own, an aspirant needs to diminish the power of pleasure and pain to bind her actions.

Much of the spare time during my teenage years was spent learning to sing. One of the exercises required by my instructor was the singing of vowels up and down the scales by thirds. The purpose of the exercise was to produce exactly the same quality sound irrespective of the pitch. In this case, under the guidance of a teacher, the natural tendency of the voice to produce different sounds at different pitches was gradually transformed into a contrived effect, whereby the voice produced somewhat more identical sounds at different pitches. Then, having been freed from the rigidity of its natural expression, my voice could be used to serve a higher purpose—bringing to life a musical composition.

This exercise in voice training is like the effort that one can make in neutralizing the tendencies of the personality. The activation of the will, which could be used to provide one with ease of living, is used instead as discipline, patience, and persistence to neutralize personality elements and enact some chosen purpose. In order more effectively to manage the difficulties brought about by the exercise of the will and the effort toward self-transformation, an aspirant can seek to integrate the elements of her personality and be open to insights originating from the transpersonal self. Let us consider these strategies, which can help to keep her from going off the road.

### Personality Integration

Occasionally wisdom greater than our normal knowledge surfaces in our experiential streams. By definition, this quality can be called transpersonal. In describing the phenomena associated with self-transformation, we should not be concerned with the particular theory that may eventually best explain them. Throughout our discussion, I shall use the schemata of scientific psychology, psychosynthesis, and theosophy wherever they can help to clarify the phenomena of interest.

As an instance of this inner wisdom, there appears to be a mechanism that provides us with information about ourselves. In particular, aspects of our subpersonality structure may be revealed. However, this information often reaches us in symbolic form, perhaps in a daydream, dream, meditation, or creative work. Then we need to interpret the spontaneously produced symbols.

We can also invoke this process, primarily in two ways. The first is to use a series of images or suggestions to elicit a response relevant to our concerns. Let us return to the example in chapter 2 for identifying subpersonalities. *Visualize a house. Use your imagination to fill in the details of its appearance. This house has a door.*

*The door opens, and someone or something walks out. Who is it? What do they look like? Are they carrying anything? Are they happy or sad? Now a second being walks out. And so on. When everyone is out of the house, have them pose for a group photograph.* When I have led such an exercise for others, I have found that the interpretation of the characters that emerge for a given person often provides her with meaningful insights into aspects of her own psychological functioning. However, there may be some reluctance initially to look at these images because there are no apparent connections among the intentions underlying the exercise, the instructions, and the subsequent imagery.

A second way of invoking the process of inner wisdom, one that provides continuity among our intentions, the strategies used, and the resultant imagery, is to pick a recurring pattern of feelings or behavior and associate an image with it. For example, at one point I wanted to characterize feelings of anxiety as a subpersonality. To do this, I recalled specific instances in which anxiety had been present and tried to identify the sensations that accompanied them. *What do I feel? I feel like a leaf being blown in the wind, having no control over my destiny.* A relevant image occurred to me. *I see a leaf, detached from the tree, swept along under heavy, threatening skies. Sadness. Hopelessness.* Thus there is a subpersonality, associated with anxiety, sadness, and hopelessness, that can be symbolized by a leaf being swept by the wind under threatening skies.

We will come back to questions about the validity of such inner images and the problem of symbol interpretation a bit later. For now, let us proceed with a description of the process of personality integration that was presented in summary form in chapter 2.

Having gained some recognition of personality elements, we can work with them, using specific exercises. In particular, if a change can be made symbolically, then this may move one in the direction of actual psychological change. We could start by refining subpersonalities in order to find their legitimate contribution to the personality. For example, in our imagination, we could walk a subpersonality up the side of a mountain, noticing the changes that are brought about with increasing elevation. In my case, I imagined the wind-blown leaf going up the side of a mountain. And what did I see at the top? *I saw myself, standing calmly and looking out across the mountain ranges and valleys. There, hovering over one shoulder, was a beautiful butterfly.* The leaf blowing in the wind had become a butterfly, still at the mercy of the wind but with some degree of control over its movement.

What did the butterfly mean? What was the positive value of anxiety? In contemplating this image, it occurred to me that fear was a mechanism through which I was warned of impending dangers. In many cases, there was good reason to be afraid. The fear served to bring danger to my attention, so that steps could be taken to deal with a situation before it became destructive. It is also noteworthy that in this idealized image, the leaf blowing in the wind did not end up being transformed so as to be a part of myself. The implication was that the fear was not actually a part of myself, even though it was associated with me. While these insights do not exhaust the psychodynamics of anxiety or solve its possible existential roots, this example does demonstrate one way in which seemingly undesirable elements of our personalities can be transformed and reconceptualized as having a role to play in our lives.

Having identified and transformed subpersonalities, one can also work with them symbolically in order to increase their cooperation. Representations of subpersonalities can be deliberately brought together and visualized as functioning in a harmonious manner. In the example of the transformation of the leaf into a butterfly, the resultant image already included the role of the subpersonality within the integrated personality.

To the degree to which a personality is integrated, it can withstand the difficulties and mitigate the dangers that can accompany spiritual aspiration. This can be illustrated using the following metaphor. When a potter creates a clay pot on a wheel, it must be as symmetrical as possible; an excess or deficiency of clay in one spot can cause the pot to fly apart, particularly if the wheel is speeded up. The personality is like the pot with excessive and deficient aspects that have to be corrected. Spiritual impulse is symbolized by speeding up the wheel. With the impact of spiritual energy, personality foibles become exaggerated and harder to contain. Just like the faulty pot, a flawed personality can break apart. A symmetrical pot, which can continue to be shaped as the wheel is speeded up, symbolizes an integrated personality, which can withstand the onslaught of stress.

One implication of this metaphor is that spirituality is exacting, and while it can be beneficial, it can also create stress for the personality. Wolff found a certain degree of physical tiredness following the presence of a transcendent current. He attributed this "intangible tiredness" to the purifying effects of the current.[9] Baker has stated that the influx of fire can destroy a person physically, emotionally, and mentally unless that fire can be expressed by working toward the resolution of planetary problems.[10] Clearly, an integrated personality is necessary in order to receive and constructively express this spiritual energy.

In other words, there is work to be done while one is waiting for enlightenment. Those who plunge in willy-nilly, hoping to experience wonderful transcendent states of consciousness, will be disappointed if they are fortunate enough not to succeed—and traumatized if they do succeed. Some of the misadventures on the spiritual path stem from the shattering of the personality by energies prematurely invoked from the transpersonal. If one hopes to function as a personality and relate constructively to one's social environment, then it is necessary to work toward personality integration.

Assagioli distinguished between personal psychosynthesis, which involves personality integration, and spiritual psychosynthesis, which is aimed at the "harmonious co-ordination and increasing unification [of the perfected personality] with the Self."[11] It is the effort to identify and encounter the transpersonal self to which we now return.

**Discernment**

In the last section, I gave examples of insights in which psychodynamic features of my personality were revealed. One can also experience meaningful spontaneous imagery that is relevant to more general concerns. Some years ago I had the following dream. *It was nighttime and I was walking near a town. My attention was drawn to an airplane, which was taking off from a nearby runway. To my horror, I realized that the airplane was too heavy to make it into the air. I watched as it plunged to the earth and crashed.* When I awoke, the dream's meaning was clear to me. I had committed myself to giving an oral presentation at an academic conference in a few weeks' time. I had a good idea of what I wanted to say but, being somewhat compulsive, had decided to write a formal paper to accompany the oral presentation. However, I was also burdened with a heavy workload and other academic commitments. I realized that I was trying to do too much. I dropped the idea of a formal paper and simply put together an outline of the material that I wanted to present. This strategy worked well when I gave the oral presentation.

The question is, What was it that alerted me? Where did the warning come from? While we could regard the dream as coincidental, let us develop the possibility that it was a warning from a wise part of myself, the transpersonal self. But that brings us back to the notion of guidance, discussed at the end of chapter 2, and the need for authenticity in order to prevent inner messages from being viewed as authoritarian demands.

In my own experience—and I do not know to what extent the conclusions that I have drawn can be generalized—one difference between the inauthentic and authentic modes of accommodating inner messages concerns passivity. In an earlier, less authentic mode, I had learned to mistrust my own understanding. Rather, I sought to passively receive an impression from the transpersonal self. Having asked a question, I would try to register whatever image first occurred to me in the hopes that the intuition could slip through before the machinery of the mind got a chance to crank out its version of an answer. Sometimes information obtained in this way would appear to be surprisingly accurate, and sometimes it would be clearly wrong. The point is that in these situations I was a passive agent whose own ideas were to be set aside in favor of an inner prompting.

In my later, more authentic mode of functioning, rather than ignoring what I knew about something, I would use that as the starting point. Rather than disregarding my own knowledge, I would seek to deepen my understanding. In reviewing what I already knew, relevant insights would sometimes occur. This process is nothing other than the concentrative style of meditation, described in chapter 4. Confining attention to the subject of interest draws forth new ideas about it. Instead of working in spite of the mind, intuition now works in cooperation with it. While this method was not as glamorous as going into a "trance" and "getting" stuff, it had the benefit of allowing me to retain my integrity.

There are, however, features common to both earlier and later modes of relating to the transpersonal self. For example, even in the later mode, silence would be valued as an internal environment conducive to deeper understanding, and insights would continue to emerge spontaneously at times when I would not be strenuously mentally engaged. The difference is that even in these more passive situations, there has been an increased degree of self-determination within the domain over which I naturally have control, which has allowed for a greater sense of integrity. These earlier and later modes are also comparable to the two ways of invoking the process of inner wisdom concerning one's subpersonalities discussed in the last section.

In talking to those who are spiritually inclined, I have often found a compliant acceptance on their part of authoritative statements, psychic pronouncements, and inner images. Discernment needs to be developed in order to sort out this plethora of information from supposed higher sources. But how is one to disentangle often contradictory guidance? How does one discern? In particular, how does one know when something is coming from the transpersonal and when it is not? Or, using the minimal definition of the transpersonal as a construct for the source of wisdom, how does one know if an inner image is an expression of wisdom or folly? Well, it may turn out that there are no universally applicable answers to these questions. But by now it should be clear that there is an obvious way to try to find out: to ask the transpersonal self.

I once participated in an eight-day intensive psychosynthesis workshop in which there was ample opportunity to work with imagery. I would ask questions, and images would arise in my mind as answers to the questions. But how did I know that the answers were true? I asked that question and got another image. *My point of view was such that I found myself outdoors on a sunny day. I could see a brick wall that was like the back wall of the church that I had attended as an adolescent, except that the brick was a dark reddish-brown and very old rather than white and new. In the image there was a narrow flower bed between the wall and the driveway at the back of the church. In the flower bed close to the wall was a small plant that had come up and was reaching toward the sun. It was boxed in on one side by the church wall and on the other three sides by glass panes. However, there was no glass pane covering the top, nor could one be placed over the top, since it would prohibit the growth of the plant.*

What kind of an answer was this? What was the meaning of the image, if it had one? I came up with the following interpretation. The church wall represented spiritual tradition. I was the little plant that had popped up outside the wall, seeking the fresh air and sunshine of truth outside the confines of religious orthodoxy. The glass panes represented certainty. While certainty about horizontal affairs—represented by the panes of glass at the sides of the plant—could protect me from a potentially destructive environment, certainty with regard to vertical matters would terminate my growth, which was symbolized by the fact that a glass pane could not be placed over the top of the plant. If I were to know without doubt that the images I was receiving were true, then I would never have to struggle to understand anything anymore. All I would have to do is follow orders. I would have become a puppet of god. And so the answer was that it was not to my benefit to know whether or not these images were true.

I also realized that this information was self-referential. Therefore I did not know whether this particular image was true or not. So I did not know whether such certainty was necessarily detrimental. However, the interpretation that I had come up with made good sense to me. Instead of being so busy trying to be told what to do, I needed to strive to deepen my own understanding and, with healthy respect for the unknown, use it as a basis for judgments and actions.

Inner images can change spontaneously as time goes by. A number of years later, I checked in on this image. The plant had grown into a strong young tree. It had managed to destroy a large part of the wall. It turned out that there was nothing behind the wall. The laterally placed panes of glass had long been shattered—just a few shards were left in the dirt. And many years later, the tree had matured. A few ruins remained of the wall. There was no sign of the glass. The earlier, brittle certainty had not withstood the impact of the development of understanding.

This is not to deny that a more resilient certitude may not eventually be possible. From a theosophical point of view, the ability to discriminate the real from the unreal is a characteristic of the buddhic nature of one's spiritual triad. But that just takes us back to where we started from—we need to undergo a process of self-transformation in order to activate this dormant quality. And in order for that process to proceed, we need to develop our own understanding. Thus the effort to develop discernment amounts to authenticity.

**Interpretation of Symbols**

Disentangling the flux of inner imagery presupposes that we can establish meanings for those images that appear to be symbolic. How are we to interpret symbols?

Some clinicians have noted that clients who go to Freudian psychotherapists see Freudian symbols in their dreams, while those who go to Jungian psychotherapists see Jungian symbols in their dreams.[12] The implication is not just that a Freudian psychotherapist would interpret the same image differently from a Jungian psychotherapist, but that psychotherapist-specific images would be dreamed by the clients. Whether or not this is true, it raises questions concerning the extent to which there can be uniform interpretations of dream symbols and the extent to which uniform interpretations are necessary.

The implication for a spiritual aspirant is that each person needs to learn to interpret her symbols using her creative imagination in the context of her understanding. A diary can be used to facilitate this process, as well as that of self-transformation more generally. The idea behind such a diary is not that it be a record of everyday events but rather a tool to assist one in the practical matter of working with oneself.[13] Even the act of writing out a dream and some of its possible interpretations can help to clarify its meaning. On the other hand, dream symbols are not entirely arbitrary, so sometimes it is helpful to solicit others' interpretations. This process can bring to light meanings that might otherwise remain unexamined.[14]

Another, more complex example involves not just my own dream imagery but that of others and further illustrates dream interpretation, suggests the possibility of sharing imagery, and speculates about possible transformations of symbols. One year, one of the students in my consciousness class wanted to carry out a study to see whether dream telepathy were possible. We set up a protocol, whereby I gave the student seven sealed envelopes, each of which contained a sheet of paper with five descriptive phrases of items or events that were to be the targets expected to show up in dreams. She would open one of the envelopes on each one of seven evenings, read the sheet of paper, and make up a story involving the five targets. Because previous studies of dream telepathy implied that strong negative emotions enhanced transmission, she would involve herself in a traumatic way in the story each evening. In addition, she would draw a picture depicting the events in the story and try to retain a vivid impression of the events as she fell asleep.[15] There were two intended recipients of the dream images: a student who believed that such transmission was possible and a student who was skeptical but willing to participate. All that the intended recipients had to do was dictate the events of their dreams into a tape recorder every morning as soon as they awoke. There was not to be any communication among the participants until the data had been gathered.[16]

Before I examined the results of the experiment, I wondered whether or not this simple study would reveal anything anomalous. In keeping with the spirit of the experiment, I had a dream. I found myself in Greenland, which, in my dream, was located in the area of the Yukon Territory. I was walking some distance from civilization, in a part of this Greenland that had remained unexplored. It had snowed. And there in the snow, to my surprise, I saw the tracks of the yeti, the abominable snowperson. In my excitement, I wanted to return to civilization and show these tracks to others, to show that the yeti really did exist. I turned around. And then I realized the predicament that I was in. My own footprints were imprinted in the snow leading up to those of the yeti. I, who had seen the tracks, was inescapably the mediator for those who had not seen them. In fact, as far as others were concerned, I could have created impressions of yeti footprints in the snow.

How did I interpret this dream? Part of Greenland appeared as an unexplored territory, which could be interpreted as the scientific study of telepathy or of anomalous phenomena more generally. Finding the tracks of the yeti is symbolic of finding the traces of anomalous events, in this case telepathy. What was thought to be myth turns out to be real. The dilemma posed by my tracks in the snow signifies the problem posed by the participation of an investigator in a scientific study. While the investigator is necessary for making the observations, if those observations are not immediately accessible to the public, then one cannot rule out the possibility that the investigator is also the source of the data.

When we analyzed the results, we were surprised at the extent of the correspondence between the stories woven around the targets and the subsequent dreams of the intended recipients. For example, one of the targets was "a broken watch." This was included in a story in which the protagonist is attacked by her husband inside her kitchen. In the ensuing struggle, she is thrown against a glass patio door, breaking her watch and cutting her right hand. In the dream of one of the recipients, she was attacked by wolves breaking into a building and chewing her hand. Upon subsequent interrogation, she revealed that it had been the right hand. Furthermore, the wolves also lunged for her head, so that she had to beat them with a "metal stick that police use." In the story that was transmitted, the protagonist used a hockey stick, one of the five targets, to strike her husband on the shoulders and head. Thus, while the research design was such that levels of significance could not be assigned to the correspondences, the results were suggestive.

What also struck us was the extent to which the initial elements of the story appeared to have undergone symbolic transmutations. For example, one of the targets was "shopping at the Bay," a large department store. Two nights after transmission, one of the recipients dreamed that she was in a car, which stopped on Huron street and then turned and continued driving on Waterloo. The first of these street names is the name of a nearby lake, and the second contains the word "water." Could it be that the Bay showed up as Huron and Waterloo? While these are both names of streets in the vicinity of our college, why did these particular streets show up and not others whose names have no thematic relationship to bays?

In fact, in some cases, the telepathic effort itself seemed to be interwoven into the fabric of the dreams. The wolves attacking the dreamer in the first case mentioned above seemed to be deliberately drawing attention to the targets. It was as though they were messengers attacking the dreamer, insisting that these features of the dream be noticed. This observation is admittedly speculative but opens up some ways of thinking about the dynamics of symbols and indicates how difficult it is to nail down empirically those phenomena that depend upon symbols.

Well, the material in this chapter has all been rather straightforward. Life is sometimes difficult, we need to persevere, we need to integrate our personalities, develop discernment, and learn to interpret symbols. It is all obvious if one is to transform oneself. Yet I am raising these points precisely because so often they fail to be taken into account. Unlike chapter 5, which consisted of speculations, this chapter has focused on the experiential level. However, it has been my observation that even where there is agreement with much of what I have said, these ideas remain at the intellectual level and fail to be carried through into practice. In fact, failing to adequately deal with the difficulties that one may encounter can have serious consequences. Let us look for a moment at what might happen to the aspirant on her way down route *C* in figure 7.

### Pathology

"Pathology" is perhaps an overly strong word to characterize the experiences of route *C* in figure 7: it implies, in no uncertain terms, that something has gone wrong. What, precisely, is meant by "pathology" here?

The discussion is not about psychopathology, although some spiritual experiences have often been considered psychopathological from a traditional point of view,[17] and even though physical pathology and psychopathology may also be present for those who are experiencing an existential crisis. Sometimes, what might ordinarily be considered pathological could be the result of experiences associated with the process of self-transformation.

For example, from the point of view of existentialism, anxiety and depression can occur when the tranquillization of inauthenticity lapses. In fact, one may be encouraged to fully open oneself to the fear created by the anticipation of one's own death as a means of breaking the stranglehold of social expectations. In these cases, depression may signify that progress is being made, with the implication that the person should not be restored to normality but should be allowed to make the transition from an inauthentic to authentic mode of being.

As a second example, in a small number of cases, a psychotic breakdown, whereby a person loses touch with the normal world, may be the result of an inability to sustain the false masks that she has been presenting for the sake of those around her. In such cases, the veneer of the persona disintegrates, and normally unconscious psychodynamics become enacted as behavior that can be interpreted symbolically.[18] Again, rather than trying to suppress the psychotic symptoms, one would seek to assist a person to learn to trust herself to be who she is.

Related to this inversion of what constitutes pathology is the distinction between pathologies that are "*progressive*," which arise from spiritual aspiration, and those that are "*regressive*," which pertain to aberrations of normal functioning in a domain where normal functioning would be expected. In fact, both types of pathologies may be present simultaneously. This distinction is a necessary one because methods that can be applied generally for the treatment of regressive pathologies may be detrimental when used for the treatment of progressive ones.[19] It is the progressive pathologies that are of interest here.

There is an odd twist to spiritual progression, I think, that creates difficulties. If there were some one thing that were guaranteed to lead to spiritual perfection, then spiritual aspiration would be a straightforward matter of doing that one thing. However, it seems to me that for a long time, it is not so much what one does but what one refrains from doing that is important. One is in a continuous process of breaking habits that are natural to the personality, which is much harder to do, since it involves mindfulness during every moment of one's life. Not doing is harder to do than doing. And so we sometimes end up doing just to do something, even though that may lead us down route *C*.

With all the emphasis in this book on personal accountability for actions, it should be pointed out that one can also become preoccupied with trying to do the right thing. This is known as "scrupulosity"—the fear of doing the wrong thing or obsessive doubt about whether or not one has done the wrong thing. Meticulousness and compulsive behaviors may also be present as one tries to regain a lost sense of purity.[20]

Scrupulosity is related to the problem of overly manipulating and assuming responsibility for everything that happens. There is a sphere of influence within which one is responsible for making decisions and over which one exercises some degree of control. The size of this sphere can fluctuate during the course of one's life and is often larger than one imagines it to be. However, it does not include everything. In particular, we live our lives within the spheres of influence of other human beings, who are also making decisions that can genuinely affect our lives. Thus we are cocreatively rather than independently responsible for some of the events in each of our lives. Consequently, even in determining the limits of decision making and responsibility, judgments have to be made.

I have been surprised by the degree to which people can grow in apparently healthy directions in spite of having participated in what I would consider to be aberrant enterprises. This applies to activities ranging from relatively mild participation in a cult to dabbling in divination that has been followed by apparent possession, hospitalization, and electroconvulsive therapy. While I do not know whether such resilience is any greater than it is in general with regard to a person's recovery from traumatic incidents, it lends credence to the notion of a self-actualizing tendency.[21]

There is a great deal of interest in spiritual matters, although much of it remains hidden from the public.[22] However, with this activity out of sight, the casualties of spiritual effort often remain unseen as well. This is particularly likely given the upbeat philosophies of many spiritual groups, which prohibit them from acknowledging that anything they do could have deleterious effects. When things go wrong, there may be little ability to adequately deal with the situation. I have seen enough spiritual casualties to realize that there is a real need for research to identify pathologies and develop ways of properly dealing with them, so I want to raise awareness of the issue and say a little bit about channeling because of its implications for knowledge and because of some of the problems that could be associated with it.[23]

**Channeling**

According to one definition, "*channeling is the communication of information to or through a physically embodied human being from a source that is said to exist on some other level or dimension of reality than the physical as we know it, and that is not from the normal mind (or self) of the channel.*"[24] However, such a definition obscures the distinction between the deepening of understanding into the superconscious on the one hand and automatism or access to low-grade material on the other. Problems can arise when the information is not "to" but "through" a human being and when the source of information is a level of reality that is not part of the superconscious.

Let me deal first with the second of these problems, the quality of channeled material. Channelers typically maintain that they are getting their information from spiritual planes or claim that to talk about planes is a misguided way of characterizing the superior source of their information. They may also be aware of a distinction that psychics sometimes make between lower and higher psychism, the former using more emotionally dependent, error-prone mechanisms of communication, and the latter using the rarely found, spiritual mechanisms. Of course, a medium, psychic, or channeler is unlikely to admit that she is using lower psychism. *Yes, all I do is rake up garbage from the lower astral planes under the influence of astral goons who have managed to sucker me into it.* Those who channel claim to be using higher psychism to channel squeaky-clean spiritual beings of superior intelligence or beings who are so superior that they do not even need to be spiritual.[25]

It should be pointed out that this superior intelligence only applies to subject areas that matter. One of the tests used to determine if someone is really channeling a being from another dimension or if she is just dredging up stuff from her own unconscious is to ask the channeled entity to add three single-digit numbers—for example, "Oh, Great Hyperborean Master, what is *2 + 4 + 5?*" If the channeled entity does not know the answer, then one can be confident that one has indeed contacted a being of superior intelligence on another plane.[26]

Well, channeling astral goons is one thing, but channeling Jesus is quite another. Or is it? Jesus is one of the most popular entities channeled as well as being a favorite topic of channeled information.[27] But does any of this material really come from Jesus? Is any of it true? Should one completely disregard it? Is it fair to lump all of it together?

To begin with, while one can make guesses about the sources of channeled information, in many cases it is not possible to eliminate various possibilities, let alone to establish which external entity is being channeled—if indeed it is possible to channel external entities. If there really are capricious astral entities, as described in the theosophical model, then there is nothing to stop them from impersonating Jesus or anything else. In practice, then, one has to evaluate the quality of the material that is received on its own merits, without being intimidated by the purported fact that it is being generated by Jesus or some other supposedly great being.[28] One way to try to do this is to shift perspective. *If I knew that the bartender had written this instead of Jesus, would I still be reading it?*

For example, does it matter who the author is of the opening quotation from chapter 1? Would it make any difference to find out that this text is supposed to be a mélange of William James and Jane Roberts?

> No more will I address my colleagues in the hallowed academic halls, for there I will have lost the credentials that once brought me such respect, since the very conditions of my existence now would make me an outcast from such conventional gatherings.
>
> In the world in which once I so gladly took my place, the dead have few rights for their existence goes unrecognized, and no psychologies prepare them for the transitions that occur as the soul moves into realms for which no earthly education can prepare it. . . .
>
> Science since my time has compromised itself, and while it was once the harbor of the truth seeker who escaped religion's dictates and dogmas, now the truth seeker must steadfastly stand apart from science and religion alike.[29]

Has the credibility of this last sentence of the quotation changed? Does it matter if it has been channeled from James? If specific channeled material makes sense in and of itself and is of practical use in dealing intelligently with issues in one's life, does it matter where it came from? Alternatively, if it does not make sense or provides no new real knowledge, then why regard it with reverence?

Having touched on the variable quality of channeled information, let us go back to consider the problem of becoming an automaton during the process of channeling. To begin with, it is important to note that the degree to which an entity blends with the channeler varies. In some cases, which appear to be benign, the channeler converses mentally with an external entity as two people would converse. An example of this would be Bailey's channeling as she has described it.[30] At the other extreme, an entity purportedly takes possession of and uses the channeler's body. Roberts has used a method of this sort.[31]

In cases where an entity largely takes over the channeler's body, the channeler may have no apparent memory of any events that transpired during the time of possession. In other cases, there is more of a "blending . . . or merging" with the channeled entity, and the channeler is "on the side" while the entity is present.[32] The whole point of this method is to get the channeler out of the way and let something else have a say.

Outside of the context of channeling, there are times when a person may feel as though she is not acting deliberately, but rather that actions are occurring effortlessly. Rather than pushing and pulling, trying to get reality to conform to one's wishes, one touches reality lightly, as it were. This is sometimes viewed as a characteristic of exceptional functioning.[33]

Perhaps a way of understanding this positive relinquishment of control is to compare it to the automation of many skills that are normally carried out. For example, when first learning to type on a keyboard, a person has to actively seek to remember where each key is located. As learning progresses, these processes become increasingly automated until the person thinks only about what she wants to say, and her fingers automatically strike the correct keys. Similarly, in learning to coordinate one's personality, emphasis is initially placed on one's personality characteristics. Increased integration may be accompanied by increased automation, so that one is less aware of personality functioning and more focused at a deeper level, so to speak, from which one allows the activity to transpire.

The distinction between channeling that involves some degree of possession and exceptional functioning could be thought of as the difference between being displaced horizontally and being displaced vertically. Vacating one's personality while something else uses it is a lateral displacement. On the other hand, exceptional functioning may imply that one has integrated one's functioning at a lower level and can direct it from a more transpersonal point of focus. One implication of this is that channeling may be an impediment to exceptional functioning, since one has been displaced from the levers of control. Disregarding for a moment variations in the nature of one's relationship to the transpersonal self that could be construed as channeling, we could say that the point is to be the source of channeled material rather than its recipient.

In channeling, the fear is that one can become possessed by an evil entity. In order to try to offset this danger, those who teach channeling may encourage people to raise their vibrations, thereby keeping out undesirable influences.[34] But how high does a person have to raise her vibrations in order to guarantee that undesirable entities will not take over her body? How does she know that these exercises in her imagination are really working and are not just wishful thinking? Having invoked Archangel Michael's protection, how does a person know that he was not otherwise engaged and actually showed up? It is one thing to believe that one is protected but quite another to know that one is.

Previously I presented the hypothesis that there may be a mechanism in reality that works to realize the contents of one's thoughts. This is the rationale behind concentrative meditation and the symbolic restructuring of the personality. Along the same lines, if such protection from undesirable entities were visualized, it would presumably move a person toward protection. On the other hand, fear of, or fascination with, evil may open a doorway to it. There is a need for balance between knowing enough about the really bad stuff so that it can be recognized and dismissed, and knowing so much that a tendency to invoke it is created. One way of dealing with undesirable influences is to refuse to play the game and to engage in constructive activities instead. Of course, one could also do that by not playing the channeling game in the first place.

Channeling can become problematic if the quality of information is poor or if a high degree of automatism is involved. In other words, if either of these conditions holds, then channeling can lead to difficulties. In particular, channeling what appears to result in good information does not mean that automatism is unproblematical. Perhaps the most famous example of such a case was Edgar Cayce, who would speak while in a sleeplike state.[35] Many people felt that they benefited from his advice. However, he did not heed warnings not to do too many readings and, it is said, died prematurely as a result.[36] It is possible that channeling involving automatism by its nature creates dangers for the personality. Those who decide that they will make that sacrifice for the sake of benefiting the world with the superior wisdom that they are channeling should be aware of the price that they may be paying and reconsider whether there may not be better ways to be of service.

Possession is a condition in which "a person's mind, body, and soul" are taken over "by an external force perceived to be a deity, spirit, demon, entity, or separate personality."[37] Undesirable cases of possession are considered to be a rare possibility.[38] However, there are some types of channeling that may be more likely to lead to undesirable possession than others. For example, particularly obnoxious entities can be accessed through the use of a Ouija board.[39] In some cases, such as that of Roberts, there can be a progression from the use of the Ouija to automatic writing and speaking.[40] There can also be episodes in which the possessed has no memory of what she has done but which may have included aggressive behavior toward others.

Possession, as well as psychological and possibly physical persecution, can occur in contexts other than those involving deliberate efforts to channel. In the worst cases, which appear to be rare, one becomes a target for those with evil intentions. Recently, through the use of hypnotic regression, cases of satanic ritual abuse have purportedly been uncovered.[41] However, because of the unreliability of hypnotic regression and the lack of corroborating evidence, the extent of criminal satanic ritual activity is unknown.[42] Whatever the degree and extent of persecution, it is a reminder to strive toward an ethical life-style, which minimizes one's involvement with potentially undesirable influences.

Throughout our discussion, we have assumed that possession by external entities actually occurs. But perhaps there are no external entities on other planes. Perhaps one simply contacts the projections of one's own mind.[43] In other words, channeling and possession may simply be the result of domination by subpersonalities that have lost their ability to cooperate with other elements of the personality. There is a form of psychopathology, dissociative identity disorder, in which a person behaves at different times as though she were distinctly different people.[44] It could be that experiences of possession are just instances of dissociative identity disorder.[45]

However, psychiatric labels are not explanations of phenomena. The opposite could also be possible—we may be using dissociative identity disorder as a diagnosis for those who are possessed by external entities. In this connection it is interesting to note that some patients with dissociative identity disorder "present personalities who claim to be spirits of deceased people."[46] Each case needs to be considered in its own right. We may end up applying the labels schizophrenia, dissociative identity disorder, possession, channeling, or hypnotic states of consciousness. Regardless of the manner in which a case is identified, one needs to make judgments about the possible presence of external entities and, where maladjustment is present, pursue a course of rehabilitation that helps to restore to the person as much integrity as possible. In other words, the psychodynamics of different cases are likely to be specific to the individual and require some degree of relevant psychotherapeutic expertise.

Finally, there is a type of disempowerment resulting from reliance on the transpersonal self that needs to be distinguished from some of the conditions that could be created by channeling. According to the theosophical tradition, at some point in their evolution, the souls of animals move out of the third kingdom, the animal kingdom, into the fourth, or human, kingdom. Their preparation for this dramatic change is to be incarnated as domestic animals. However, animals who become domesticated cannot subsequently survive in the wild. Having become dependent upon human care, they lose the skills necessary to find, kill, and compete for food. Similarly, movement from the fourth kingdom to the fifth, the kingdom of souls, is accompanied by decreased ability to function in the human kingdom without the support of the "masters" of the fifth kingdom carrying out the "domestication." This can be discussed in the context of social adjustment following an existential crisis depicted in figure 7. The new levels of adjustment associated with route *A* are the result not of the reestablishment of old ways of survival but rather of new ways that depend, in part, on the transpersonal self. For example, with a growing sense of compassion, one may cease to compete with others for educational and employment opportunities and rely instead on one's intuition and synchronistic occurrences. Thus self-transformation, as described in this chapter, is an irrevocable process that can lead to the disempowerment of the personality and reliance on the transpersonal self.

CHAPTER SEVEN

# Conclusion

## *The Convergence of Science and Spiritual Aspiration*

> [Gnostic realization] is, in a sense, neither religion nor science as ordinarily understood, and yet combines features belonging to each.
>
> —*Franklin Wolff*

How far have we got in examining our efforts to know the meaning of life and the nature of reality, the elephant we have been trying to make sense of? What can we say about the creature and the groping?

We start by being lost in inauthenticity. Our interpretations of events and our actions are unknowingly determined by the schemata that make up our belief systems. From a theosophical point of view, we are controlled by exemplars on the astral plane that govern the patterns of events on the etheric and physical planes. In turn, our thoughts unknowingly strengthen these schemata, thereby contributing to the limits of knowledge and behavior for our culture. In this way our lives are organized by a rigid system of adult valuing.

Materialism and scientism are features of an inauthentic mode of being. In keeping with our naive interpretations of everyday experience, we think of our world as ultimately made up of tiny, well-behaved billiard balls. We assume that we encounter concrete objects, blithely interpret phenomena in terms of our everyday perceptions, and remain convinced that we will end up in oblivion when we die. Scientism, the handmaiden of materialism, precludes the possibility of deeper understanding of reality by placing an emphasis on the accumulation of incontrovertible facts and restricting the methods and content of investigation. Thus consciousness is thought to be an emergent property of the brain, existence to be meaningless, and spirituality to be a delusion for those too weak to face the facts.

Then something happens. Maybe our physical existence is threatened, or we see an unidentified flying object on our way home from work, or our everyday lives make us nauseous. We start to wonder what is going on. But the schemata that are supposed to be answers to the existential questions no longer satisfy us. They are too thin, too superficial. And then we realize that we do not know the answers and that neither does anyone around us, even though they pretend to know. And so we start looking. We become black sheep, then goats scrambling on the cliffs of our inner consciousness. Our families may reject us, our friends think that we are weird, and our colleagues tell us that we will "end up selling candles on the beaches in Southern California."[1]

Perhaps we shop in the spiritual marketplace for answers to our questions. We may get turned on to colon irrigation or raising kundalini or channeled guidance from Jesus. We are searching, looking for answers. In some cases, our approach to reality may not have changed, with one set of schemata simply having been substituted for another. Our ability to retain credibility within society may completely evaporate as we cease to be able to earn a living or carry on a conversation without trying to convert our listeners. We may end up in a psychiatric ward, struggling with inner demons that have been loosed in the process of inquiry and are running amok inside of us.

It is not easy. We may find that we are enmeshed in webs of schemata. The harder we struggle, the more we get tangled. And then slowly, perhaps, we develop a disciplined but gentle approach that allows us to disidentify our selves from the confusion and to create a space for something deeper to become manifest. As we start to deliberately develop our understanding, schemata may become flexible and change from being barriers to becoming tools.

We may also realize that we need to reorganize our lives. We need to integrate and eventually synthesize the disparate elements of our personalities. This is not a process that occurs automatically but a practice that requires deliberate, persistent effort. This goal is consistent with the emphasis in the Western religious and mystical traditions on the need for an ethical life-style and service to humanity. For Frankl, meaning ensues when we transcend our personal concerns in favor of expressing love, working on a worthwhile project, and responding constructively to the challenge of unavoidable suffering. There is no shortage of need for goodwill in the world today. Social injustice, environmental degradation, the threat of nuclear wars and accidents, gratuitous violence, and ignorance can all be mitigated by the constructive activity of those who aspire to knowledge.

Thus far, I have tried to characterize the crises and discipline that can counteract inauthenticity. They consist in movement away from social influence. Like a spacecraft leaving the earth, we need enough power to escape the earth's gravitational field. Once in space, however, we may start to be drawn by the gravitational pull of other celestial bodies. So it is with our quest for knowledge. However, while we can gradually identify what it is that we are leaving behind, we often do not understand what it is that we are moving toward.

Historically, a number of different spiritual disciplines have been used as a propellant for this journey, including witnessing, concentrative, and reflexive styles of meditation. In authentic science, the same cognitive mechanisms appear to be engaged as those in concentrative meditation; for example, seeking to comprehend the results of experiments in quantum physics may lead a physicist to recognize the limitations of materialist schemata and cause her to seek out more comprehensive ways of understanding reality. Or perhaps, as Wolff has suggested, mathematics can contribute toward the activation of introception, which allows us to know in another way in a transcendent state of consciousness.

There are intimations of a deeper level of reality that can become apparent to us. We may begin to have meaningful, precognitive dreams. We may start to experience a greater sense of compassion toward others. Or we may realize that we are involved in a series of magically beneficent synchronous events. It is as though there were a hidden dimension within ourselves, more intelligent than we are, that occasionally irrupts in our lives. We have called this the superconscious or the spiritual triad. But how do we bring this aspect of consciousness and reality into focus? What is our relationship to the unknown spiritual source? What happens to our self-identification?

We find that we need to learn to interpret the symbols that occur during dreams or moments of reverie. We need to develop discernment so that insights, revelations, and intuitions can become integrated into our understanding. And we can allow transpersonal fluid to infuse our arid personalities. We may yet become lost. We may end up channeling or following guidance to the point where we are nothing but a puppet of god. Authenticity, the effort to act on the basis of our own understanding, which is so important for departure from the default interpretations provided by society, continues to serve us well when we are attracted to the unknown.

At some point, perhaps, we experience events described as a first initiation in the theosophical literature. Personal ambitions lose their allure. Questions of meaning become paramount. A commitment is made to seek the root. And one becomes suspended between two worlds. There is no turning back to a life of meaningless mediocrity. However, the transcendent is but a vision; it may yet dissipate along with the effortless techniques, Yaqui teachers, and astral goons. But there is an important point that is difficult to grasp for those who are not likewise suspended: the everyday world also disappears, just as a dream dissipates when we awaken. Furthermore, it is not just that it has ceased to exist; it has ceased to have ever existed.[2] Having dispelled the illusion that a fixed set of schemata delimit reality, there is nothing to which to return. There is no choice but to continue.

And where do we end up? I do not know. Our discussions of authenticity and science in chapters 2 and 3 could be said to belong to the past, that from which we emerge, the context for any discussion concerning the nature of consciousness and reality. Self-transformation—the dynamics of the development of understanding—which we discussed in chapter 6, belongs to the present. For the future, the reader has to go back to our discussions of transcendence and theory in chapters 4 and 5 and make up her own mind about the possibility of transcendent states of consciousness.

Our point of departure in the first chapter was a quotation purportedly from James, channeled by Roberts, to the effect that both science and religion must be discarded by the truth seeker. If we seriously want to answer the fundamental questions, then we need to free ourselves from the schemata of materialism and scientism. And if it is knowledge that we are striving for, then we need to discard blindly held religious beliefs. Knowledge is the currency of science, and when it is freed from the black-and-white constraints of its inauthentic aspects, science can be opened up to a colorful exploration of consciousness and the ultimate nature of reality. Then a more open, possibly introspective scientific venture may be indistinguishable from the efforts of a spiritual aspirant seeking to deepen her understanding of reality. This is the point of convergence intimated by Wolff in the opening quotation of this chapter. Having looked underneath the surface of both science and spiritual aspiration, we have found the same thing: the need for authentic knowing.

There. That's one wrinkle. Now, what happened to the rest of the elephant?

# GLOSSARY

The definitions of the following terms are the result of my own understanding of them, taking into account the usual connotations of these terms, their uses by specific authors, and their interplay in the text of this book. These are definitions, not statements about the nature of reality. That is to say, my effort here has been to identify the concepts to which specific terms refer, not to indicate the degree to which these concepts may coincide with reality. In those cases where specific sources have directly given rise to a definition, they have been identified in the notes for the text or the glossary.

**absorption.** The capacity that a person has for becoming preoccupied with the object of her attention.

**adult valuing.** Term used by Carl Rogers to refer to a state of the valuing process in which one's preferences are based on the norms of one's community.

**aesthetic continuum.** Term used by Franklin Wolff to refer to perception prior to conceptualization.

**aikido.** Japanese martial art derived from jiu-jitsu in which harmony with the spirit of the universe is emphasized.

**alta major center.** The higher throat center in the etheric body.

**altered state of consciousness.** State of consciousness in which the quality of subjective experiences differs from the normal waking state.

**ambiguity.** English translation of a term used by Martin Heidegger to denote the uncertainty as to whether or not that which we hear has been understood through a genuine effort as well as the tendency to surmise and pass on the surmising as that which is actually happening.

**anomalous.** Outside the explanatory scope of the traditional scientific worldview.

**archival analysis.** Research strategy in the social sciences in which information already on record is analyzed.

**Assagioli's egg diagram.** Roberto Assagioli's diagram of the human psyche.

**astral.** Pertaining to the plane of consciousness and reality predominantly characterized by emotionality.

**astral travel.** Travel in the astral plane.

**astrology.** Art of relating celestial events to human affairs.

**atmic.** Pertaining to the level of consciousness and reality characterized predominantly by transpersonal will.

**attention.** Bringing to awareness.

**authentic.** That which is true to its own nature.

**authenticity.** Effort to act on the basis of one's own understanding.

**authentic science.** Effort to acquire knowledge about reality through open-minded investigation.

**autogenic training.** Method involving the imagination of bodily sensations, used to voluntarily control autonomic nervous system activity.

**automatic writing.** Writing by a person who is not in conscious control of that which is being written.

**awareness.** Cognizance.

**behaviorism.** Movement in psychology concerned with observable human and animal behavior.

**belief system.** A person's constellation of schemata and scripts.

**buddhic.** Pertaining to the level of consciousness and reality characterized predominantly by intuition.

**celestial equator.** Projection of the earth's equator onto the background of fixed stars.

**channeling.** The production of information or energy that is alleged not to originate with the person producing it.

**collective unconscious.** Unconscious domain shared by all human beings.

**concentrative meditation.** Meditation in which the aspirant attends to a single object of thought.

**connectionism.** Characterization of cognition in terms of interconnected units.

**consciousness.** Subjective stream of thoughts, feelings, and sensations, some of which are more directly the focus of attention than others; and a sense of existence and efficacy.

**constructivism.** View that mystical experience is significantly shaped and formed by the subject's beliefs, concepts, and expectations.

**correlational study.** Method of research in the social sciences in which the strength of the relationships between variables is determined.

**curiosity.** English translation of a term used by Martin Heidegger to denote our propensity for seeking novelty rather than coming to know entities.

**differentiated aesthetic continuum.** Term used by Franklin Wolff to refer to the realm of ordinary perceptual experience.

**differentiated theoretical continuum.** Term used by Franklin Wolff to refer to the rational conceptualization of reality.

**disidentification.** Term used by Roberto Assagioli to refer to the relinquishment of the illusion that we are a particular aspect of our personality.

**dissociative identity disorder.** Condition in which a number of distinct personalities appear to be present in an individual.

**ecliptic.** Path traced by the sun against the background of fixed stars in the course of one year as viewed from the earth.

**electroencephalogram.** Recording of the electrical activity emanating from the brain.

**electro-oculogram.** Recording of the electrical potentials of eye muscles.

**emergent property.** Property of a system that results from the activity of lower levels of the system and is dissimilar from the properties of the lower levels.

**enlightenment.** Realization of the truth.

**esoteric.** Inner or hidden.

**etheric.** That which forms the substrate for the physical.

**etheric centers.** Centers of energy in the etheric body.

**etheric tracts.** Pathways of energy in the etheric body.

**evenly suspended attention.** Attitude on the part of the therapist in traditional psychoanalytic psychotherapy characterized by the suspension of rational thinking and impartial attention to everything there is to observe.

**existential crisis.** Crisis precipitated by concerns about one's existence.

**existentialism.** Movement in philosophy in which efforts to resolve existential questions have been grounded in human experience.

**existential questions.** Fundamental questions concerning existence.

**extrapersonal.** That which is mundane but beyond the limits of one's personality.

**fifth kingdom.** Spiritual kingdom of souls and angels.

**fourth kingdom.** Human kingdom.

**gnostic realization.** Esoteric, spiritual knowledge.

**guided imagery.** Sequence of suggested images.

**guru.** Spiritual teacher and possibly conveyor of spiritual power and liberation.

**hatha yoga.** Yoga through the use of physical disciplines.

**higher mental.** In the theosophical theory, pertaining to the level of consciousness and reality characterized predominantly by abstract functioning of the mind.

**higher psychism.** Accurate methods of obtaining information from unseen dimensions of reality.

**higher self.** Term used by Roberto Assagioli to refer to a person's true, though usually unrealized, identity.

**homeopathy.** System of healing based on the use of highly diluted substances to treat those symptoms of a disease that would be produced by the substance in healthy individuals.

**horizontal.** Dimension of reality at the level of the personality.

**humanistic psychology.** Movement in psychology that emphasizes human experience.

**human potential movement.** Social movement characterized by the use of diverse methods to help individuals develop their latent potential.

**hyperbola.** A pair of curves in a plane whose points have the characteristic that the difference of the distances from any point on one of the two curves to two fixed points in the plane is a constant.

**ida.** Etheric tract related to consciousness.

**idle talk.** English translation of a term used by Martin Heidegger to denote a person's transmission of unexamined information from the environment to others.

**imperience.** Term introduced by Franklin Wolff to refer to awareness that is the analogue of experience in a transcendent domain.

**inauthenticity.** State of being in which one is not true to oneself.

**indeterminate aesthetic continuum.** The transcendent ground of perceptual experience in Franklin Wolff's philosophy.

**indeterminate theoretic continuum.** Term used by Franklin Wolff to refer to the transcendent ground of knowledge.

**infant's way of valuing.** Term used by Carl Rogers to denote a continuous preconceptual evaluation and selection of the elements of one's experience with regard to the degree to which they tend to actualize one's organism.

**information.** Organized distinctions.

**initiation.** Expansion of consciousness in the course of one's spiritual development.

**insight.** Realizations concerning the nature of mind that occur in witnessing meditation.

**instrumental transcommunication.** Phenomenon of apparent communication with the dead through electronic means.

**interference pattern.** Pattern that results from the interaction of waves.

**introception.** Term introduced by Franklin Wolff to refer to the cognitive faculty, distinct from sense perception and conceptual cognition, that provides knowledge of transcendent reality through identity.

**introjections.** Material from outside oneself embodied in one's psyche.

**introspection.** Subjective examination of the contents and processes of consciousness.

**intuition.** Error-prone irrational knowledge or infallible revelation of truth.

**judgmental heuristics.** Mental shortcuts in which simple schemata are used for evaluating complex situations.

**karmic retribution.** Circumstances resulting from past misdeeds.

**kriya yoga.** Union of oneself with transcendent reality through a specific action.

**kundalini.** Latent energy of matter residing in the base of spine etheric center.

**locus of control.** Internal to external range of the source of control of a person's ideas and actions.

**low-balling.** Compliance technique in which an initial incentive to perform an action is removed after the victim has established other reasons for that action.

**lower mental.** Pertaining to the level of consciousness and reality characterized by the ordinary functioning of the mind.

**lower psychism.** Error-prone methods of obtaining information from unseen dimensions of reality.

**mantra.** Word or phrase used for specific purposes other than communication.

**materialism.** The belief that everything that exists is material in nature.

**materialist-transcendentalist continuum.** Continuum of beliefs about consciousness and reality that ranges from the belief that reality is material in nature to the belief that reality is grounded in transcendent states of consciousness.

**mathematical yoga.** Term introduced by Franklin Wolff to refer to the union of oneself with transcendent reality using mathematics.

**mature way of valuing.** Term used by Carl Rogers to denote the valuing process in which both cognitive evaluation and the wisdom of one's organism serve to enhance the individual.

**meditation.** The effort to transcend the mind.

**mindfulness.** The process of attending to, labeling, and dismissing thoughts in witnessing meditation.

**monad.** Spiritual aspect of an individual's psyche, considered to be a cell in the body of God.

**monadic.** Pertaining to the most spiritual level of consciousness and reality of which we can become conscious.

**multivariate statistical procedures.** Statistical methods used for the simultaneous examination of the relationships between a number of variables.

**mystical experience.** Experience characterized by enlightenment and exceptional emotional well-being.

**near-death experience.** Experience reported by a person to have occurred while she was close to dying or believed to be dead.

**neurotransmitter.** Chemical used by nerve cells for communication.

**new age.** Social movement characterized by an interest in the paranormal and an eclectic approach to spirituality; or referring to the movement of the vernal equinox into the constellation Aquarius.

**noetic.** Quality of the mind pertaining to knowing.

**normal distribution.** Specific bell-shaped probability distribution with the greatest likelihood for events around the mean and the least likelihood for events furthest away from the mean.

**occult.** Hidden knowledge concerning the supernatural.

**ontic experiences.** Peter Nelson's term for experiences attributed to the transcendent.

**operational factors.** Peter Nelson's term for the actions that precipitate preternatural experiences.

**paranormal.** Outside the range of that which is normal.

**participant observation.** Research strategy in the social sciences whereby an investigator makes clandestine observations as a participant in the activities of a group that is of interest to her.

**perennial philosophy.** School of thought characterized by the idea that there is a single spiritual reality underlying life and that this spiritual reality has found expression in primitive mythologies and the various religions of the world.

**personality.** Totality of the mundane aspects of a person.

**phenomenological.** Pertaining to experience as it presents itself.

**pingala.** Etheric tract for the energy that feeds matter.

**pluralistic ignorance.** Failure on the part of everyone in a group to recognize an emergency resulting from the misjudgment that the lack of concern on the part of others identifies the situation as a nonemergency.

**ponderability.** Term introduced by Franklin Wolff to refer to that attribute of a phenomenon that gives it the appearance of being real for ordinary consciousness.

**possession.** Condition in which a person has been taken over by an external entity.

**prana.** Energy stemming from the first subplane of each plane of consciousness and reality.

**precession.** The wobble of the rotational axis of the earth relative to the fixed stars.

**precognitive dream.** Dream that includes events that subsequently occur.

**preconscious.** Material of which one is not conscious but which is readily accessible to consciousness.

**preternatural experiences.** Peter Nelson's term for anomalous experiences.

**probability distribution.** The likelihoods for specific events.

**progressive relaxation.** Method of relaxation whereby one systematically decreases tension in skeletal muscle groups.

**psychic.** Pertaining generally to the subjective aspects of personality or to paranormal experiences.

**psychoanalysis.** Psychodynamic theory and psychotherapy developed by Sigmund Freud in which conflicts resulting from childhood sexual and aggressive impulses are emphasized.

**psychodynamic.** That which is characterized by the notion that psychological functioning is analogous to fluid dynamics.

**psychomotor epilepsy.** Condition in which automatic movements are produced by disturbed neural activity.

**psychosynthesis.** Psychodynamic theory and psychotherapy developed by Roberto Assagioli in which the will, the synthesis of personality elements, and the reorientation of an individual's self-identity from the personality to the higher self are emphasized.

**psychotic breakdown.** Serious breakdown in psychological functioning that may involve disorganization of behavior and the presence of perceptions and beliefs that have no correspondence in reality.

**pure mathematics.** Mathematics that is unconcerned with the development of applications.

**rainbow bridge.** Pathway for consciousness constructed between the personality and spiritual triad.

**reductionism.** Assumption that a complex system can be entirely explained in terms of the functioning of simpler, lower-level components.

**reflexive meditation.** Meditation in which the aspirant seeks the subject for whom there are objects of consciousness.

**relaxation response.** Restful pattern of physiological activity.

**remote mental healing.** Healing brought about through the mental activity of a distant healer.

**schema.** Collection of information about a potential object of attention.

**science.** Cultural movement concerned with the acquisition of knowledge about reality.

**scientific method.** Idealized set of procedures involving the use of one's senses for making systematic observations of objective events and drawing inferences from them using one's rational faculties in order to establish general laws.

**scientism.** Inauthentic science, characterized by unexamined belief in materialism and the scientific method.

**script.** Cognitive representation of a typical sequence of events in a particular situation.

**self.** The subject of consciousness.

**self-actualization.** The process of becoming all that one can become.

**serotonin.** The neurotransmitter 5-hydroxytryptamine.

**shamanism.** Practice in some tribal cultures in which the practitioner—the shaman—enters an altered state of consciousness, possibly travels in other realms of being, and uses the resources of the altered state for the benefit of the community.

**social cognition.** Within social psychology, the study of noticing, interpreting, remembering, and using information about the social world.

**social psychology.** Subdiscipline of scientific psychology concerned with the social dimensions of cognition and behavior.

**soul.** Spiritual aspect of a human being.

**spirit.** Impulse that is the cause of all manifestation.

**spiritual.** Pertaining to religious experience or the transcendent rather than the mundane.

**spiritual triad.** Configuration of higher mental, buddhic, and atmic bodies.

**standard deviation.** Measure of the variability of a set of scores around their average.

**subatomic events.** Physical events involving phenomena smaller in size than an atom.

**subconscious.** Processes and material pertaining to the personality of which one is not conscious.

**subpersonality.** Term used by Roberto Assagioli to denote part of an unintegrated complex of psychological material in a person.

**substantiality.** Term introduced by Franklin Wolff to refer to the presence of transcendent reality in awareness.

**superconscious.** Transcendent aspect of oneself of which one is not normally conscious.

**sushumna.** Etheric tract related to spirit.

**sympathetic nervous system.** Part of the nervous system involved in the expenditure of the energy of an organism.

**synchronicity.** Acausal connectedness of meaningful coincidences.

**synesthesia.** Perception of stimuli in senses other than the one in which the stimuli are presented.

**t'ai chi ch'uan.** Chinese martial art based on Taoist philosophy.

**talking white.** Saying only those things that are acceptable to the social group to which one belongs.

**tarot.** Deck of 78 cards, often used for divination, consisting of 22 major cards and 56 minor cards divided into 4 equal suits.

**temporal lobes.** Lower middle part of the cortex of the brain.

**theoretical continuum.** Term used by Franklin Wolff to refer to the noetic dimension of consciousness.

**Theosophical Society.** Society founded in 1875 by Helena Blavatsky and Henry Olcott to study occult topics.

**theosophy.** Any system of thought based on direct experience of the divine.

**therapeutic touch.** Form of treatment in which a healer uses her hands to restore a patient's balance of energy.

**thread.** Line of energy emanating from the monad that connects it to the spiritual triad and personality.

**transcendence.** The surpassing of limitations, such as those of human experience.

**transcendental meditation.** Type of meditation, taught by Maharishi Mahesh Yogi.

**transcriptive thought.** Term introduced by Franklin Wolff to refer to the process of thinking and resultant formulations of thought in which conceptions are containers for introception.

**transpersonal.** That which transcends the personality.

**transpersonal self.** That aspect of the self which transcends the personality.

**understanding.** Meaningful knowing.

**valuing process.** Carl Rogers's term for the process by which we establish preferences for certain objects and objectives over others.

**vertical.** Mundane/transcendent dimension of reality.

**will.** Roberto Assagioli's term for a multidimensional, intimate aspect of the self whose function is to decide what is to be done, to apply the necessary means for its realization, and to persist in the face of difficulties.

**witnessing meditation.** Meditation in which the aspirant observes the stream of consciousness.

**yoga.** Sanskrit term meaning union of oneself with transcendent reality.

# *Selected Biographies*

**Assagioli, Roberto (1888–1974),** also known as Roberto Grego Assagioli. He was interested in medicine, education, and religion; received a medical degree from the University of Florence, Italy, in 1910, where he wrote his dissertation about psychoanalysis; and subsequently trained with Eugen Bleuler in Zurich. He was close to Alice Bailey and other Theosophists. Assagioli developed psychosynthesis, a theory of the psyche and a form of psychotherapy. In 1926, he founded a training center and publishing house, the Instituto di Cultura e Terapia Psichica, and, in 1961, the Instituto di Psicosintesi. He was a participant in the development of humanistic and transpersonal psychology in the 1960s. Assagioli has published about three hundred papers, and his books include *Psychosynthesis* and *The Act of Will* (Assagioli 1965, 1973; Hardy 1987; Greening 1975).

**Bailey, Alice Anne (1880–1949).** She was born La Trobe-Bateman in Manchester, England. Her work for the Young Women's Christian Association took her to India. Subsequently she moved to California, where she became a member of the Theosophical Society and, in 1917, the editor of the society's periodical. Bailey has claimed to have heard in 1919 the voice of a Tibetan lama who requested her assistance with some writing. Their purported collaboration resulted in the publication of nineteen books, including *A Treatise on White Magic* and *Esoteric Healing*. In 1923 she and her husband founded the Arcane School to disseminate spiritual teachings (Bailey 1951a, 1953, 1951c; Stephenson 1983; Shepard 1991, 1:150–51).

**Baker, Douglas M. (1922– ).** He was born in London, England, but raised in Natal, Republic of South Africa. He has learned about paranormal phenomena from the Zulus and Hindus of South Africa, the writings of Alice Bailey, and his own extensive experiences. He graduated in medicine from Sheffield University in 1964 and subsequently was the medical advisor to the de la Warr Laboratories. Baker has organized ongoing International Festivals of Esoteric Science and lectures extensively, both to British and foreign audiences. He has written one hundred books, including *The Theory* *and Practice of Meditation, Life after Death in a Nuclear War,* and *Baker's Dictionary of Astrology for the Twenty-first Century* (Baker 1975c, 1982, 1991, personal communication, 1995).

**Rogers, Carl Ransom (1902–87).** Rogers broke away from traditional religious attitudes as a result of attendance at a six-month World Student Christian Federation Conference in China and later studies at the Union Theological Seminary. He received graduate degrees from Teachers College at Columbia University, helped to establish the Rochester Guidance Center in 1939, and taught at the Ohio State University, the University of Chicago, and the University of Wisconsin. He was the founder and resident fellow of the Center for Studies of the Person in California from 1963 until his death. Rogers is known for having developed a client-centered approach to psychotherapy, participating in the founding of humanistic psychology and leading encounter groups. His books include *On Becoming a Person* and *A Way of Being* (Goleman 1987; Kirschenbaum & Henderson 1989; Moritz 1962, 11–13; Rogers 1961, 1980).

**Walsh, Roger N. (1946– ).** He graduated from the University of Queensland with degrees in psychology, neurophysiology, and medicine and is currently a professor of psychiatry, philosophy, and anthropology at the University of California at Irvine. His approximately two hundred publications on science, philosophy, medicine, religion, and ecology include sixteen books, among them *The Spirit of Shamanism; Paths beyond Ego: The Transpersonal Vision,* edited with Frances Vaughan; and *Meditation: Classic and Contemporary Perspectives,* edited with Deane H. Shapiro, Jr. Walsh is also an associate editor of the multivolume *Encyclopedia of Psychology.* His work has received more than a dozen national and international awards (Walsh 1990, personal communication, 1995; Walsh & Vaughan 1993b; Shapiro & Walsh 1984; Corsini 1994).

**Wolff, Franklin Fowler (1887–1985),** also known as Franklin Merrell-Wolff. He graduated with a bachelor's degree in mathematics from Stanford University in 1911. Subsequently he studied philosophy at Harvard University and taught mathematics at Stanford University for one year each before retiring from the world to lead a contemplative life. He experienced two fundamental realizations in 1936, which formed the basis for his philosophy. Wolff is the author of four hundred hours of taped lectures, reflections, and conversations; numerous unpublished papers; and a number of books, including *Pathways Through to Space* and *The Philosophy of Con-*

*sciousness without an Object* (Merrell-Wolff 1994, 1995b; D. Leonard, personal communication, 1993–94; R. Leonard, personal communcation, 1994–95).

## NOTES

### Chapter One: Introduction

The epigraph for this chapter is from Roberts 1978, 113.

1. Feminine personal pronouns are used in this book for generic cases as a counterbalance to the historical use of masculine pronouns.

2. Goring (1994, 527) defines theosophy as "any system of philosophical or theological thought based on the direct and immediate experience of the divine," and points out that the term has been used, in particular, to denote "the principles of the Theosophical Society."

3. Myers 1984, 17.

4. Thorngate 1990, 264.

5. Phenomenological methods have been described in Lindauer 1994 and Keen 1975. The movement in philosophy known as phenomenology has been characterized, for example, in Blackburn 1994, 284–85.

6. The new age has been discussed in Sebald 1984; Albanese 1993; Swets & Bjork 1990; Vitz & Modesti 1993; Joy 1985; Guiley 1991, 403–7.

### Chapter Two: Authenticity

The ideas in this chapter were first presented on February 12, 1993, in a seminar that was part of the King's College Faculty Seminar Series. I am grateful for comments made by participants at the seminar. The epigraph for this chapter is from Bogart 1992, 20.

1. Existentialism has been characterized by Giorgi 1994.

2. The following descriptions of idle talk, curiosity, and ambiguity have been summarized from Heidegger 1962, 211–19.

3. My purpose here is to demonstrate the continuity of the fundamental dynamics of cults with those of society at large. Some of the same issues that are raised here have been explicitly addressed in Kramer & Alstad 1993 in the context of a discussion of authoritarianism.

4. Definition adapted from Spilka, Hood, & Gorsuch 1985, 250.

5. It has, however, been argued—for example, in Haden 1989, 12; and Mansbach 1991, 66—that complete extrication from the inauthentic condition is not possible.

6. In general, humanistic psychology developed independently of European existentialism, according to DeCarvalho (1990, 35–36).

7. The summary of Rogers's ideas in this section has been taken from Rogers 1967b.

8. Latané & Darley 1968.

9. Cialdini 1988, 110.

10. Baron & Byrne 1991, 122.

11. Ashforth & Fried 1988, 306; cf. Sutherland 1989, 391.

12. Cialdini 1988, 7.

13. Baars 1988; Cialdini 1988.

14. Pallak, Cook, & Sullivan 1980.

15. The term "locus of control" is used in a literal sense here. Phares (1994) gives a discussion of its meaning in psychology.

16. Formally, Heidegger did not value authenticity over inauthenticity, although it is hard to avoid that conclusion when reading *Being and Time* (1962). Similarly, scientific psychologists generally do not judge the merits of a situation but simply describe its dynamics. However, much of social psychology, which initially developed in order to understand tragic situations, such as Nazism and the failure of bystanders to intervene in emergency situations, was based on values about our collective welfare (Blanchet et al. 1992). Rogers and humanistic psychologists highly value individuation and the development of one's potential (Shaffer 1978).

17. Beehler 1990, 40.

18. Rogers 1967a, 1.

19. These passages are found in Stevens 1970, 3; Stevens 1984, 36; and Stevens 1984, 145, respectively.

20. The relationship between authenticity and morality has been discussed, for example, by Haden 1989.

21. Aron 1977.

22. Geller 1982.

23. Murphy 1992b, 36, 32.

24. This was not the last time advances in engineering provided metaphors for psychological functioning. Behaviorists modeled psychological processes using a telephone switchboard, and computationalism arose with the invention of computers. More recently, the limitations of such borrowing have been pointed out in Wilber 1982, in the case of holographic models of the mind, which resulted from the invention of lasers.

25. Hardy 1987, 13.

26. Assagioli 1991, 30.

27. This diagram has been adapted from Assagioli 1933–34, 188; and Assagioli 1991, 30. Similar diagrams can be found in Assagioli 1965, 17; Ferrucci 1982, 44; and Hardy 1987, 23. My explanation of the diagram is based primarily on these five sources.

28. Assagioli 1965, 18.

29. A discussion of the collective unconscious has been given in Combs 1993. Unlike Jung, Assagioli thought that the collective unconscious should be differentiated so that archetypes that are archaic in nature do not get confused with those that are spiritual (Hardy 1987, 51–56).

30. Assagioli acknowledged that he got the idea from William James's heterogeneous personality (James 1958, 140–56), who in turn drew on similar ideas of earlier thinkers (Hardy 1987, 34–37).

31. Assagioli 1965, 74–77.

32. Hardy 1987, 57.

33. Ferrucci 1982.

34. Ferrucci 1982, 49.

35. Assagioli 1973.

36. Assagioli 1973, 6.

37. According to Bailey 1951a, 23, spirit is that "impulse . . . which is the cause of all manifestation."

38. For an example of hearing voices, see Caddy 1987. Intuition has been discussed more generally by Vaughan 1979a; Cosier & Aplin 1982; Silverman 1983; Hill 1987; Bowers, Regehr, Balthazard, & Parker 1990. About the I Ching, see Wilhelm & Baynes 1967; about tarot, see Graves 1973; Campbell & Roberts 1979.

### Chapter Three: Science

The epigraph for this chapter represents a reaction from a mechanistic point of view to the possibility of parapsychological events quoted by Combs & Holland (1990, xx), who discuss such attitudes. Targ & Puthoff (1977, 169) have recounted a similar reaction to their work concerning extrasensory perception.

1. These three views are presented in James 1904; D. B. Klein 1984, 175; and Chang 1978, 113, respectively.

2. This work has been reported in Baruss 1990a; Baruss & Moore 1989, 1992; and included in summary form in Baruss 1992.

3. The exact number is not known; 1,043 paper copies were mailed out. The remaining copies were sent out electronically to the recipients of the *Psychnet Newsletter*.

4. Twenty-six psychologists who indicated areas of specialty in experimental, animal, or physiological psychology or statistics were included as part of the 109 respondents in the natural and applied sciences. The 163 respondents in the social and medical sciences included 3 respondents who indicated their disciplinary affiliation as law and 2 affiliated with physical education. The 16 respondents indicating their disciplinary affiliation as transpersonal psychology were included as part of the 61 respondents associated with the humanities. The total number of respondents in the three general disciplinary categories only adds up to 333, since the disciplinary affiliation of one respondent could not be determined.

5. These included hierarchical cluster analyses, factor analyses, and bivariate analyses using the chi-square test of statistical significance. The details have been given in Baruss 1990a, chapter 5.

6. I am quoting from items Q9, Q31, Q33, and Q47 from the consciousness questionnaire (Baruss 1990a, 121, 120, 125, 124).

7. At that point the factors were turned into scales and the data were reanalyzed in order to determine the psychometric properties of the scales (Baruss & Moore 1992).

8. This dimension has been noted by others such as Frank 1977; Harman 1987; and Osborne 1981. Similar dimensions had been found empirically by Coan 1968; Krasner & Houts 1984; and Kimble 1984 for the psychological community.

9. These four views are expressed in Massaro & Cowan 1993, 385; Crick & Koch 1992, 153, and Hofstadter 1979, 709–10; Churchland 1980, 207; and Lycan 1987, respectively. My presentation is somewhat of an oversimplification of a scientific approach to consciousness with which not all researchers tending toward a materialist position would agree. However, the net result would be the same in that they would believe that consciousness is ultimately derived from physical processes.

10. Blanpied 1969; Sudbery 1986.

11. Hanson 1963, particularly 556–57. In this book, the word "material" refers to the use of the billiard-ball schema, or something like it, as a descriptor of the fundamental nature of reality and the term "physical" refers to the spectrum of experience normally labeled "physical," irrespective of how that is ultimately explained.

12. This is an unsystematic observation discussed in Baruss 1993. However, while the persistence of the billiard-ball schema in the face of contrary information is accessible to empirical investigation, neither the

presence of such a schema nor its persistence in the face of contrary evidence has been systematically studied.

13. The following description of the two-slit experiment has been adapted from Feynman, Leighton, & Sands 1965; Sudbery 1986; and, with some modifications for the single-slit case, Kafatos & Nadeau 1990.

14. Wheeler 1983; Kafatos & Nadeau 1990, 46–47; Clark 1992.

15. Kafatos & Nadeau 1990, 66.

16. This is known as the Einstein-Podolsky-Rosen paradox, first proposed as a thought experiment in Einstein, Podolsky, & Rosen 1983. This particular account is adapted from Kafatos & Nadeau 1990, 62–72.

17. Kafatos & Nadeau 1990; Aspect, Dalibard, & Roger 1982.

18. Quoted from an interview with Zajonc by Clark 1992, 22, edited by Zajonc.

19. There have been extensive discussions about possible relationships between the mental and the subatomic, for example, in Wigner 1983; Walker 1970, 1977; Stapp 1985; Elitzur 1989; Kafatos & Nadeau 1990.

20. Walker 1970, 1977.

21. Bauer 1992, 137.

22. Sloan 1992, and Tart 1992, have discussed the Western mindset within which these specific beliefs can be situated.

23. Cf. Allen 1994.

24. Bauer 1992, 32.

25. Bauer 1992, 144.

26. Zimmerman 1984, 175.

27. This refers to the eighteenth-century disbelief in meteorites, discussed in Westrum 1992, 1.

28. Similarly, in describing Niels Bohr's defense of the predicted results of the Einstein-Podolsky-Rosen paradox against Einstein's disbelief in quantum theory, Rosenfeld (1983, 143) has stated that "it is necessary to renounce any pretension to impose upon nature our own preconceived notion of what 'elements of reality' ought to be, and humbly take guidance, as Bohr exhorts us to do, in what we can learn from nature herself."

29. This story has been adapted, with some modifications, including a gender reversal, from Ornstein 1972, 187.

30. There are distinctions between subjectivity and privacy. For example, the experience that a woman may have of giving birth to a child is an experience that is privately available to her and must be inferred by others through observations of physiological changes and behavior. But

there are also qualia, the noninferential perceptions of what it is like to have a child, that are subjective aspects of her experience. Whether or not such qualia exist and, if they do, how one is to account for them, has been a source of difficulty in the philosophy of mind (Dennett 1988; Lycan 1987).

31. However, Natsoulas (1992, 287) has stated that he believes that someday it would be possible for one person to have "immediate, noninferential awareness of another person's mental-occurrence instances by means of electrodes, and so on, that connect the two people's mind-brains in a suitable way."

32. Bickhard has discussed some of the "Misconceptions of Science in Contemporary Psychology" (1992, 321).

33. Maslow 1966b; Shaffer 1978.

34. Bauer 1992, 144; Pelletier 1985, 248; Tart 1992, 78.

35. Boulting 1993, 19.

36. Josephson & Rubik 1992, 15, 16.

37. Mandler 1985, 56.

38. Dennett 1978, 173.

39. Clearly, one can also be fooled in that one may be convinced that one knows when in fact one does not (Lyons 1986; White 1982).

40. Lycan 1987, 121. Lycan's arguments for a materialist account of consciousness have serious shortcomings, as I have argued in my review of his book *Consciousness* (Baruss 1990b).

41. This observation has been made in Abelson 1988.

42. Olson & Zanna 1993.

43. Cult member quoted in Festinger, Riecken, & Schachter 1956, 168.

44. Anderson & Kellam 1992.

45. In a factor analysis of responses to belief items, Robert P. Abelson found three factors relevant to conviction: "emotional commitment," "ego preoccupation," and "cognitive elaboration" (Abelson 1988, 273).

46. I have avoided saying that the "paradigm changes," because Kuhn (1970, 175) has acknowledged that he has used the word "paradigm" with two different meanings.

47. Baruss 1990a, 123.

48. Bauer 1992, 47–51, 143.

49. Kukla 1983.

50. Pekala 1991, 27.

51. These situations are exemplified in Dennett 1978, 173, and described in Moody 1988, 67, respectively.

52. Logan 1993.

53. There would be many ways in which different beliefs could be used for making discriminations among groups of scientists so that there would be a whole class of diagrams of this sort. Moreover, the same belief may end up on a number of hills.

54. See Castaneda 1971a, 1971b, 1972, 1974 for an account of some of his adventures. The legitimacy of these accounts has been questioned in De Mille 1990. Churchill (1992, 216) has argued that when genuine natives try to correct misrepresentations of native spirituality created by nonnative "experts," such as Castaneda, they are rejected as not being Indian enough. Kremer (1992) has argued that Donner 1991, and Abelar 1992, which are about similar adventures involving some of the same characters found in Castaneda's writing, strengthen his case. This illustrates the point that the degree of access to observations is a variable, as is one's reliance on the perceived integrity of the investigators.

55. De Mille 1990.

56. Haisch 1993, 107.

57. Some of those who received a copy of the questionnaire but disagreed with a transcendentalist point of view may have been less likely to respond to the questionnaire because it included items concerning transcendental beliefs. However, this did not seem to deter other respondents with strong materialist beliefs, who appeared to use the questionnaire as an opportunity to express their disdain for transcendentalism.

58. A scientist could, after evaluating evidence available to her, conclude that all of reality could be accounted for in terms of physical events. The point is not that she agree with transcendental beliefs, which in and of themselves are neither authentic nor inauthentic, but that there be an effort to move away from rigid categories toward an understanding grounded in genuine investigation.

59. As reported in Baruss & Moore 1992, we found that in every case where there were statistically significant differences between male and female respondents, women tended toward transcendentalism. Because academia is largely male-dominated, it was hard to find women for participation in the study and difficult to know how to interpret the results. However, it seemed reasonable to use a male figure to represent a scientistic point of view in this story.

60. On extrasensory perception, see Tart 1992; Bem & Honorton 1994; Krippner & George 1986. On hydrocephalic children, see Lewin 1980; Lorber 1965, 1968; Lorber & Zachary 1968. On out-of-body experiences,

see Alvarado 1989; Mitchell 1981; Blackmore 1982. On synchronicity, see Combs & Holland 1990; Peat 1987; Keutzer 1984. On mind-body interactions, see Ader & Cohen 1993; Schneider et al. 1990; Myers & Benson 1992. On mental healing, see Byrd 1988; Dossey 1989, 1991; Miller 1982. On reincarnation, see Stevenson 1993; Stevenson, Pasricha, & Samararatne 1988; Pasricha 1992.

61. This work has been reported in Jahn & Dunne 1987; Dunne, Jahn, & Nelson 1985; Nelson et al. 1991; Jahn, Dunne, & Nelson 1987. This account is based on the above sources as well as demonstrations by Roger Nelson at King's College, University of Western Ontario, in May 1990, and participation by myself and some of my students at the Princeton Engineering Anomalies Research Laboratory, February 24–26, 1993.

62. The initial project was done by an undergraduate student using a random number generator on a computer (Fishman 1990, 46).

63. Dunne, Jahn, & Nelson 1985, 4.

64. Jahn & Dunne 1987, 96.

65. Nelson et al. 1991, 16.

66. This effect has been found at various sizes of the database. In Nelson et al. 1991, the database was generated by 108 individual operators over 11 years and included 5.7 million trials with each trial consisting of 200 pulses (2–3). This database included pulses generated not only by random-event generators but also by electronic and algorithmic pseudorandom-event generators (9–12). Using a regression model, the effect of intention was statistically significant at $p<.00024$ with an effect size of one pulse per 10,000 (15).

67. Nelson et al. 1991, 15, 16.

68. There was a consistent trend toward positive pulses that was not statistically significant (Nelson et al. 1991, 16).

69. Jahn & Dunne 1987, 124–35.

70. The level of statistical significance of the overall deviation from expectation is $p<.000003$, although only the intention to move the balls to the left relative to the baseline distribution is independently statistically significant (Jahn, Dunne, & Nelson 1987, 37).

71. Near-death experiences have been discussed by Moody 1988; Morse 1990, 1992; Schröter-Kunhardt 1993; Moody & Mishlove 1988; Bauer 1985; Bates & Stanley 1985; Carr 1993; Wren-Lewis 1988.

72. Moody 1988, 5.

73. Morse 1992, 24.

74. Morse 1992, 61–83; Moody 1988, 35.

75. Morse 1992, 86.

76. Compared to 4 percent of normal adults who have not had near-death or paranormal experiences (Morse 1992, 132).

77. Morse 1992, 132–33.

78. Malamud 1986 discusses the degree of experienced reality in a comparison of lucid dreaming with the normal waking state.

79. Gescheider 1985, 3.

80. Fuller 1985; Macy 1993; and Harsch-Fischbach 1989, 1992.

81. Fuller 1985, 10, 11; Macy 1993, 17.

82. Macy 1993.

83. Macy 1993, 18; Harsch-Fischbach 1992, 38–39.

84. As gathered from the descriptions given by Harsch-Fischbach 1989 and 1992.

85. Harsch-Fischbach 1992, 28.

86. Macy 1993, 17. Some of the history of contacts using human channels has been discussed by Klimo 1987.

87. Harsch-Fischbach 1992, 28.

88. Macy 1993, 19.

89. For example, Osborne 1981; Reason & Rowan 1981; Messer 1985; Leighton 1990.

90. See Battista 1978; Pekala 1991; Reason & Rowan 1981, for the incorporation of subjective aspects of experiences; Keen 1975; Barrell et al. 1987; and Pekala & Levine 1981, for the advocation of phenomenological methods; and Keen 1975, 41, for the contention that phenomena can be better revealed using phenomenological methods than they can in ordinary experience.

91. Concerning the discreditation of introspection, see Lyons 1986; and White 1988. Mandler and Dennett, as discussed in the previous section, have tacitly accepted it.

92. For example, in the form of thought and experience sampling (Singer & Kolligian 1987).

93. Needleman 1965, 98.

94. Concerning participation by anthropologists in native cultures, see Turner 1992, 28; concerning the measurement problem in physics, see Wheeler & Zurek 1983; concerning the induction of phenomena that one is interested in studying in consciousness, see Walsh & Vaughan 1980, 173; Blackmore 1982, 106; Dane & DeGood 1987, 98; concerning the methodology used by Wilhelm Wundt, see Apfelbaum 1992, 534.

95. Belief in life after death as a result of interacting with those who have had near-death experiences has been exemplified by Moody in

Moody & Mishlove 1988; increased appreciation for life as a result of interacting with those who have had near-death experiences is discussed in Morse 1992, 77, 78. Quotations are from Ring 1993, 3.

96. Lilly 1978.

97. Minton 1992, 548.

98. Something along the lines of a noosphere (Teilhard de Chardin 1959, 182; Goudge 1967), implicate order (Bohm 1980), or morphogenetic field (Sheldrake 1981, 1988).

99. Statements against the use of more liberal methodologies are made in Hilgard 1980; Natsoulas 1978b. The quotation is from Hilgard 1980, 15.

#### Chapter Four: Transcendence

1. Dawson 1987, particularly 242–44.
2. Maslow 1971, 269.
3. Frankl 1984, 133; 1966; 1984.
4. Frankl 1966, 99.
5. Frankl 1984, 141.
6. Ornstein 1972, ix.
7. Walsh & Vaughan 1993a, 7.
8. Maslow 1966a, 1968.
9. Quotation from Helena Petrovna Blavatsky's *The Voice of the Silence* taken from Baker 1977b, 14.
10. Other overviews of meditation have been given in West 1987; and Shapiro & Walsh 1984. The three categories given here are consistent with those given in Ferrucci 1990. Other authors have used other schemata for categorization. For example, Goleman (1988) has described two main categories, essentially the first two given here, as well as a number of specific techniques.
11. Murphy & Donovan 1983, 1988.
12. The derivation of transcendental meditation from Shankara is in Goleman (1988, 66) and is implied by Russell (1976, 22–25). Shankara—also referred to as Shankaracharya (for example, in Goleman 1988, 66)—resuscitated Hindu philosophy in India (Shankara 1947). In this system, the goal of spiritual practice is a state of nonduality (Goleman 1988). Recent and contemporary proponents of Shankara's philosophy have included Ramaṇa Maharshi (Mahadevan 1977), his disciple, H. L. Poonja (Ingram 1992), and his disciple, Andrew Cohen (1992). On Maharishi Mahesh Yogi, see Guiley 1991, 620; Russell 1976, 26.

13. At least, that was my experience of it. Such a loose characterization of the technique could be criticized by proponents of transcendental meditation, such as Russell (1976), who would claim that the purity of the technique must be preserved in order to experience the benefits of transcendental meditation.

14. The information about transcendental meditation mantras in this paragraph has been taken from Goleman 1988, 67; Scott 1978, 45–54; and Morris 1984, 129.

15. Blood lactate results from skeletal muscle metabolism. Increased levels have sometimes been associated with the onset of anxiety attacks (Benson 1975, 65–67).

16. Benson 1975, 112–15; 1977; and 1983, 283–84.

17. Carrington 1986, 497.

18. Epstein & Lieff 1981, 138.

19. Younger, Adriance, & Berger 1975.

20. Cauthen & Prymak 1977; Holmes 1987; Michaels, Huber, & McCann 1976; J. C. Smith 1976.

21. Farthing 1992, 438.

22. J. C. Smith 1976, 634.

23. For example, in Crider et al. 1993, 195.

24. Pelletier 1985, 148. Another counterexample has been given in Holmes 1987, 102.

25. Concerning clinical uses, see Carrington 1987; concerning military uses, see Heckler 1990.

26. Goleman 1988.

27. Epstein 1984. The first two quotations are from Freud, quoted in Epstein 1984.

28. Goleman 1988. A classic Hindu text describing concentrative meditation is the *Yoga Sutras* of Patanjali (Ferrucci 1990, 113; Goleman 1988, 71).

29. Ferrucci 1982, 109.

30. Baker 1975c.

31. Bailey 1950a, 95–96.

32. Wallas 1926; Harman & Rheingold 1984.

33. Natsoulas 1978a; and 1986, in response to Helminiak 1984.

34. Lyons 1986.

35. Spilka, Hood, & Gorsuch 1985, 176–77.

36. Spilka, Hood, & Gorsuch 1985, 182.

37. Greeley 1987, 8.

38. Baruss 1990a, 169.

39. Thomas & Cooper 1980, 78, 79. The percentages given here are those in Thomas & Cooper 1980.

40. Green & Green 1986, 557–58; and Assagioli 1991, 82, respectively.

41. Walsh et al. 1980, 41–42.

42. Allman et al. 1992, 564.

43. Caird 1987; Spanos & Moretti 1988. However, Spanos & Moretti (1988) found that diabolical experiences are correlated with psychopathology.

44. These characteristics are cited in Greeley 1987, 8; Lester, Thinschmidt, & Trautman 1987; and Thomas & Cooper 1980, 81, respectively.

45. Poloma & Pendleton 1991, 74, 76.

46. Spilka, Hood, & Gorsuch 1985, 192.

47. Spilka, Hood, & Gorsuch 1985, 193, 196 (quotation).

48. Nelson 1989, 193, 194.

49. Nelson 1990, 39.

50. Operational factors "can be understood as the immediate set of methods, procedures, and activities which, against the appropriate personality background, directly trigger the conditions necessary for the production of a given praeternatural experience" (Nelson 1990, 38); Nelson 1990, 40, 42.

51. These descriptions are approximations of those given in Nelson 1990, 42, 47.

52. Nelson 1989, 198.

53. Nelson 1989, 195.

54. J. C. Smith 1987, 141.

55. Nelson 1990, 42, 43.

56. Mandell 1980. The quotations are from pp. 400, 436. An attributional model of religious experiences more generally is developed in Spilka, Hood, & Gorsuch 1985.

57. Forman 1990, 3.

58. Wilber 1975, 1979, 1980; Maslow 1964, 1968, 1971.

59. Murphy 1992a.

60. This evidence has been gathered from biographical accounts of extraordinary individuals, adult recall of childhood religious experiences, and the reports of children (Armstrong 1984, 209).

61. Armstrong 1984, 223. The meanings of "soul" have been discussed in Eliade 1987.

62. Moore and I found that 13 percent of respondents agreed that "reincarnation actually does occur" and that only 58 percent disagreed (Baruss 1990a, 174).

63. For example, Armstrong 1984; Maxwell & Tschudin 1990.

64. This is evidenced by the fact that the items "I have had an experience which could best be described as a transcendent or mystical experience" and "My ideas about life have changed dramatically in the past" had factor loadings of .767 and .529 respectively on the same factor (Baruss 1990a, 123).

65. Ring 1987, 174, 175. Similarly, Morse (1992) has argued that the vital aspect of near-death experiences is their spiritual nature.

66. The quotations in this sentence are from Walsh 1976, 101. All other quotations and material concerning Walsh's experiences have been taken from Walsh 1984.

67. Baker 1975c.

68. Baker 1977b, 5, 6.

69. Baker 1977a.

70. The information concerning Wolff and his ideas in this section and the next has been based on material given in the reference list as being authored by Wolff, Merrell-Wolff, Ron Leonard, and Rowe, as well as unpublished material by Dorothy Leonard. For some of his writing, Wolff added his first wife's maiden name, so that the name of the author would appear as "Franklin Merrell-Wolff." I am following R. Leonard 1991 in referring to Wolff by his legal name, Franklin Fowler Wolff.

71. Merrell-Wolff 1975.

72. Merrell-Wolff 1973b, 19, 30; 1970, 187. Wolff has distinguished between "liberation" and "enlightenment," with the first fundamental realization corresponding to liberation and the second to enlightenment (Merrell-Wolff 1973a, 278). In Merrell-Wolff 1975, he maintained that while the former was in part the result of his efforts, the latter was spontaneous and unexpected.

73. Merrell-Wolff 1973b, 24.

74. Merrell-Wolff 1970, 181.

75. Merrell-Wolff 1973b, 35–36.

76. Merrell-Wolff 1970, 198; 1973c, 36–37; 1973a, 277.

77. Merrell-Wolff 1973b, 38–55.

78. Merrell-Wolff 1973b, 42; 1975.

79. Quotations are taken from Merrell-Wolff 1975; 1973a, 277–78. See also 1973c.

80. Merrell-Wolff 1975, 1973a. I have replaced Wolff's word "felicity" with "facility."

81. Merrell-Wolff 1970, 138.

82. The description in the following two paragraphs is a reconstruction of relevant passages taken largely from Merrell-Wolff 1970, 137–203.

83. Merrell-Wolff 1973b, 96. Ron Leonard (1991, 258) has attributed the use of the term "transcriptive thought" to Wolff.

84. Merrell-Wolff 1966b, 1966c. Ron Leonard (1991, 149–50) has remarked that Wolff could have been more explicit about his mathematical yoga.

85. Merrell-Wolff 1966f. Only one of the two curves of a hyperbola is shown in figure 3.

86. Merrell-Wolff 1966f.

87. Merrell-Wolff 1966f.

88. This hyperbolic function is continuous as long as its domain is restricted to the positive real numbers. If it were defined on the set of all real numbers, it would be symmetric about the origin, with a discontinuity at zero.

89. Merrell-Wolff 1966e. In the sixth sentence of the quotation, the word used by Wolff may have been "above" rather than "about."

90. Merrell-Wolff 1966a. As an example of an infinite series, Wolff uses $1 + 1/2 + 1/4 + 1/8 + \ldots$, which adds up to 2.

91. Wolff 1939, 414.

92. Merrell-Wolff 1966c.

93. Northrop 1966; Merrell-Wolff 1966e.

94. Merrell-Wolff 1966e; Wolff's comments concerning his own study of mathematics are based on unpublished material obtained from Doroethy Leonard.

95. Merrell-Wolff 1966d; 1970, 143.

96. Merrell-Wolff 1966d; 1970, 159.

97. Merrell-Wolff 1966a. According to Merrell-Wolff 1966a, in effect, there is no sacrifice. After contact with the transcendent, everything is returned, although one's role has become that of steward. If something were not returned, then that would be just as well, since one would be better off without it.

98. Merrell-Wolff 1966b.

99. Assagioli and Wolff downplayed this influence in their public presentations. However, Assagioli and Wolff's first wife were associates of

the theosophist Alice Bailey, and in both cases, theosophy appeared to have played an important part in the development of their own understanding.

### Chapter Five: Theory

The epigraph for this chapter comes from Fadiman 1992, 212. Fadiman has asserted that Murphy 1992a has not done this in his book *The Future of the Body: Explorations into the Further Evolution of Human Nature.*

1. Mishlove 1993. For example, Young 1976a, 1976b.
2. Campbell 1980, viii; see also Lacombe 1982.
3. Regis 1987; Green, Schwarz, & Witten 1987.
4. Smolensky 1988; Hanson & Burr 1990.
5. Bruce 1972.
6. Information concerning the origins of the Theosophical Society has been taken from Gomes 1987; Campbell 1980; and Hutch 1980.
7. Bailey 1951c, 133, 137; Campbell 1980, 151.
8. Campbell 1980, 151. Adyar, India, had become the headquarters for the faction of the Theosophical Society eventually led by Annie Besant and C. W. Leadbeater after Blavatsky's death (Guiley 1991, 613–14).
9. Bailey 1951c, 158. Other details concerning her membership in the society can be found in this work.
10. Campbell 1980, 152.
11. Bailey 1951c, 167.
12. These books have been described in *The Work of the Master Djwhal Khul with Alice A. Bailey* ([Bailey] 1974). Purported dates of transmission for them are given in Stephenson 1983, 23. These books are listed in the references with A. A. Bailey as the author, along with those that she did not claim to have written with the Tibetan.
13. Baker 1977b, 5.
14. Bailey herself acknowledged this ambiguity in 1951c, 168.
15. Bailey has encouraged the reader to "ponder"—for example, in Bailey 1972 and 1955a—and Baker has recommended "brooding" in Baker 1977b, 68–80. And as Bailey said, "Students must remember that the aim of all truly occult teachers is not to give information but to train their pupils in the use of thought energy" (Bailey 1962a, 868).
16. There have been other sources for my understanding of theosophical ideas, including Besant 1897; Blavatsky 1962, 1971; Leadbeater 1988, 1915; Chevalier, 1976; and Spangler, 1975, 1977a, 1977b, 1980, 1984.
17. This diagram has been adapted from Baker 1975c.
18. These subplanes and planes are not spatially layered "like the shelves of a book-case" (Leadbeater 1915, 17).

19. Including quarks (Phillips 1980).

20. Blavatsky 1962, 2:286; Stephenson 1983.

21. This drawing has been adapted from Baker 1975c.

22. A definition of the thread is given in Bailey 1962a, 114.

23. These lines are known as "nadi" in Sanskrit. See Motoyama 1978, 85–86. An overview of Chinese medicine is given in Eisenberg 1990.

24. One of the functions of etheric bodies is the same as that of the morphogenetic fields proposed by Rupert Sheldrake (1981, 1988), namely, to produce specific forms for physical manifestation.

25. Motoyama 1978, 118.

26. Hatha yoga is described in Guiley 1991. According to Bailey, (1955b, x, 121), it is laya yoga that is concerned with the etheric body, while hatha yoga is restricted to the physical.

27. On t'ai chi, see B. Klein 1984. On aikido, see Heckler 1990; Guiley 1991.

28. Ming-Dao 1983.

29. Krieger 1981.

30. Ullman 1988.

31. Davenas et al. 1988; Benveniste 1993; O'Regan 1988.

32. The Sanskrit word "chakra," denoting a wheel or disc (Blavatsky 1971, 75) is commonly used to refer to them.

33. Adapted from Bailey 1953, 715; Baker 1976b, 59; Baker & Hansen 1977, 169.

34. Bailey 1953, 185. Cf. Ossoff 1993; Sannella 1976.

35. Bailey 1953, 176, 170.

36. Bailey 1953, 156, 160, 161.

37. Bailey 1953, 151, 153.

38. James 1904.

39. The information concerning James's links to Eastern religions was provided by Eugene Taylor. It is important to note that James was well aware of neural activity, as evidenced, for example, in chapter 9 of *The Principles of Psychology* (1983, 248), where he has given a description of "neurosis underlying consciousness." Nonetheless, he chose to explain thinking in terms of breathing.

40. The information in this paragraph about the higher throat center has been taken from Baker 1975c.

41. Bailey 1953, 147, 145.

42. Baker 1975c.

43. Bailey 1953, 171.

44. Shamanism is discussed, for example, in Walsh 1990; and Picknett 1990, 98–99.

45. Tzu 1968, 1, 66.

46. Baker 1976b, 59.

47. Bailey 1953, 184.

48. Bailey 1953, 185.

49. For example, Bailey 1953, 185–86.

50. I will use the words "material," "matter," and "substance" to refer to the substrate within which astral phenomena occur, even though these terms no longer have all the connotations that are associated with them at the everyday level of experience. Bailey (1950b, 208) said that "the astral plane in reality does not exist." However, to put this in perspective, "the dense physical substance is *not* a principle" (Bailey 1960b, 377). According to Baker, astral phenomena can be encountered objectively.

51. Baker 1982, 9.

52. This process is described in Bailey 1951a.

53. I have identified these dimensions from discussions of lucid dreaming, such as that in LaBerge & Gackenbach 1986.

54. Baker 1982, 13, describes lower and higher astral bodies, a distinction that has not been made explicitly here.

55. This definition is consistent with the discussion in Malamud 1986.

56. Bailey 1985, 34.

57. Moody 1988, 8–9.

58. Bailey 1985, 37–38.

59. Moody 1988.

60. Bailey 1951a, 40–41; Bailey 1955b, 253.

61. Morse 1992, 196.

62. Moody 1988; Morse 1992.

63. This problem is also raised in Blackmore 1982, 26.

64. Farthing 1992; LaBerge & Gackenbach 1986.

65. Academic discussions include Alvarado 1989. General discussions include Harary & Weintraub 1989; Mishlove 1993; Mitchell 1981; Fox 1962; Monroe 1972; Yram 1967.

66. Blackmore 1984; Levitan & LaBerge 1991.

67. The tarot is discussed, for example, in Melton, Clark, & Kelly 1991, 129–35; Graves 1973; Campbell & Roberts 1979; Guiley 1991. The tarot can be used as a cosmology independent of divination. On automatic writing, see Guiley 1991, 45–46.

68. The association of spirituality with ascent does not preclude its association with descent—for example, into a deep well.

69. Bailey 1965, 27. Nine have been discussed in Bailey 1960b.

70. Bailey 1960b, 666.

71. These characteristics of the first initiation have been taken largely from a public lecture by Douglas Baker on November 23, 1975, in Toronto, Ontario.

72. These characteristics of the first initiation have been taken from a public lecture by Douglas Baker on November 23, 1975, in Toronto, Ontario; and Bailey 1960b, 667.

73. These characteristics of an initiate have been taken from a public lecture given by Douglas Baker on April 6, 1975, in Toronto, Ontario.

74. Bailey 1960b, 683.

75. These characteristics of the second initiation have been summarized from Bailey 1960b, 678–79; and a public lecture given by Douglas Baker on November 23, 1975, in Toronto, Ontario.

76. Bailey 1960b, 684.

77. Bailey 1960b, 44.

78. The material in this paragraph concerning the third initiation has been taken from Bailey 1960b.

79. Mishlove 1993 has an overview of the research concerning astrology. Nienhuys 1993 and Ertel 1993 express contrary points of view concerning possible evidence for the correlation of specific astrological and human events.

80. Based on R. Baker & Fredrick 1971, 47.

81. Metz 1974 and Hieratic 1975 are examples of astrologers' ephemeri.

82. Robson 1976, 16; R. Baker & Fredrick 1971, 50.

83. The notions of emotional refinement and lateral penetration have been taken from Baker 1975c.

### CHAPTER SIX: SELF-TRANSFORMATION

1. This is true of the impacts of stressors more generally (Rodin & Salovey 1989, 557).

2. This possibility would be consistent with the stages of spiritual development described in Helminiak 1987. An account of Helminiak's theory is reviewed in Rayburn 1993.

3. This would also be considered normality as opposed to psychopathology, as defined, for example, in Davison & Neale 1986.

4. Cited from Ferrucci 1982, 165; and Bailey 1955a, 132, respectively.

5. Some styles of therapy never get beyond this point. Rogers, who originated client-centered therapy, would simply provide a supportive environment and allow a person to sort herself out (Rogers 1961).

6. This can be seen from the survey provided in Crider et al. 1993. What makes something pleasurable is a much harder question to answer (e.g., Wise & Rompre 1989).

7. Bibby 1987.

8. Kornfield 1993, 34.

9. Merrell-Wolff 1973a, 6.

10. Baker 1977a, 79.

11. Assagioli 1965, 30.

12. Vaughan 1979b, 104.

13. Baker has been more specific about the uses of a spiritual diary, for example in 1977b.

14. A protocol for doing this is described in Ullman 1986.

15. Technically, were she to have succeeded, this would not necessarily have been dream telepathy, since the student was trying to send only from a waking state, and the recipients may not have been asleep at the time that she was sending.

16. This study is briefly described in Johnson 1990.

17. Walsh et al. 1980, 41–42.

18. The direct interpretation of schizophrenic behavior using psychoanalytic constructs is discussed in Rosen 1953 and, using more general interpretations, in Dusen 1967, although the degree to which the language of schizophrenics is symbolically meaningful has been questioned (Maher 1979).

19. Assagioli 1991, 129–30.

20. Spilka, Hood, & Gorsuch 1985; Greenberg, Witztum, & Pisante 1987.

21. Abraham H. Maslow's use of the term "self-actualization"—which is close to its use in this book—to mean spiritual self-transformation was derived from Kurt Goldstein's utilization of the term to mean an organism's spontaneous reorganization incorporating damages from injuries (DeCarvalho 1991, 120).

22. There are formal secret societies, some of which purportedly also have political and financial power (Kinney 1988; Kinney & O'Neill 1989; Delaforge 1988; McIntosh 1988, 1989; Wilson 1988). However, there are also less secretive networks of people who share common spiritual interests.

23. A general discussion of spiritual problems is given in Bragdon 1993.

24. Klimo 1987, 2.

25. There is also "open channeling," in which the source of channeled material does not become identified (Klimo 1987, 304).

26. Hughes 1991, 172.

27. For example, *The Aquarian Gospel of Jesus the Christ*, channeled by Levi H. Dowling; *The Urantia Book*, published by Urantia Foundation, whose means of transmission is unknown according to Guiley (1991, 630–33); the newsletter *Pearls of Wisdom*, channeled by Elizabeth Clare Prophet (my comments are based on copies of issues from 1975, when they were published by The Summit Lighthouse, and from 1978, when the copyright holder was given as Church Universal and Triumphant, Inc.); a series of mailings, each of which was entitled "OPEN LETTER to Speedily Advancing Persons," sent out by "Ruby Focus" of Magnificent Consummation, Inc., produced by Garman and Evangeline van Polen (my comments are based on copies of letters from 1976 and 1977); *A Course in Miracles*, channeled by Helen Cohn Schucman according to Klimo (1987, 37–42), and published by Foundation for Inner Peace; and Benjamin Creme's *The Reappearance of the Christ and the Masters of the Wisdom.*

28. This solution does not work entirely. Part of the point of channeled material is that the source is in a dimension normally inaccessible to us, so that presumably we receive insight into a level of reality to which we ordinarily do not have access (Anderson 1988, 8). If that is the focus of our interest, we would need to have some degree of confidence about the source of the material.

29. Roberts 1978, 21, 113.

30. Bailey 1951c, 163–64.

31. Roberts 1978, 12–20.

32. Hughes 1991, 166–67, 173.

33. Maslow 1968, 106.

34. Hughes 1991, 167–68.

35. Stearn 1989.

36. Guiley 1991, 84.

37. Guiley 1991, 457.

38. Perry 1990, 4.

39. Guiley 1991, 418–19; Klimo 1987, 198; Perry 1990, 2.

40. Guiley 1991, 419.

41. Ryder 1992.

42. Persinger 1992; Clifton 1989; Benson et al. 1991; Fadiman 1993.

43. This would be consistent with a Jungian analysis (Corey 1988, 86).

44. American Psychiatric Association 1994, 484–87; O'Regan 1985; Watkins & Watkins 1986. This condition was previously known as "multiple personality disorder."

45. Richeport 1992.

46. Braude 1988, 192.

### Chapter Seven: Conclusion

The epigraph for this chapter is from Merrell-Wolff 1970, 202.

1. Walsh 1984, 49.
2. This point is made more generally in Merrell-Wolff 1966c.

### Glossary

Archival data. Tiemann 1993.

Attention. Novak 1987.

Autogenic training. Raimy 1994.

Behaviorism. Cheney 1993.

Enlightenment. Defined in Goring 1994, 160.

Evenly suspended attention. Epstein 1984; Giovacchini 1987, 256–57.

Guru. Defined in Goring 1994, 202–3; Guiley 1991, 249–50.

Humanistic psychology. Defined in M. B. Smith 1994.

Hyperbola. Thomas 1968, 344–51.

Kriya yoga. Yogananda 1981.

Perennial philosophy. Huxley 1946.

Prana. Defined in Baker 1975c.

Theosophy. Sykes 1976, 1201; Gomes 1987, 87.

Understanding. Defined in Blackburn 1994, 386.

Yoga. Mann 1994.

# REFERENCES

Abelar, T. 1992. *The Sorcerers' Crossing.* New York: Viking Arcana.

Abelson, R. P. 1988. "Conviction." *American Psychologist* 43, no. 4: 267–75.

Ader, R., & N. Cohen. 1993. "Psychoneuroimmunology: Conditioning and Stress." *Annual Review of Psychology* 44:53–85.

Albanese, C. L. 1993. "Fisher Kings and Public Places: The Old New Age in the 1990s." *Annals of the American Academy of Political and Social Science* 527:131–43.

Allen, M. J. 1994. "Scientific Method." In *Encyclopedia of Psychology*, edited by R. J. Corsini, 3:355–57. 2d ed. New York: John Wiley & Sons.

Allman, L. S., O. de la Rocha, D. N. Elkins, & R. S. Weathers. 1992. "Psychotherapists' Attitudes toward Clients Reporting Mystical Experiences." *Psychotherapy* 29, no. 4: 564–69.

Alvarado, C. S. 1989. "Trends in the Study of Out-of-Body Experiences: An Overview of Developments since the Nineteenth Century." *Journal of Scientific Exploration* 3, no. 1: 27–42.

American Psychiatric Association. 1994. *Diagnostic and Statistical Manual of Mental Disorders.* 4th ed. Washington, D.C.: Author.

Anderson, C. A., & K. L. Kellam. 1992. "Belief Perseverance, Biased Assimilation, and Covariation Detection: The Effects of Hypothetical Social Theories and New Data." *Personality and Social Psychology Bulletin* 18, no. 5: 555–65.

Anderson, R. 1988. "Channeling." *Parapsychology Review* 19:6–9.

Apfelbaum, E. 1992. "Some Teachings from the History of Social Psychology." *Canadian Psychology/Psychologie canadienne* 33, no. 3: 529–39.

Armstrong, T. 1984. "Transpersonal Experience in Childhood." *Journal of Transpersonal Psychology* 16, no. 2: 207–30.

Aron, A. 1977. "Maslow's Other Child." *Journal of Humanistic Psychology* 17, no. 2: 9–24.

Ashforth, B. E., & Y. Fried. 1988. "The Mindlessness of Organizational Behaviors." *Human Relations* 41, no. 4: 305–29.

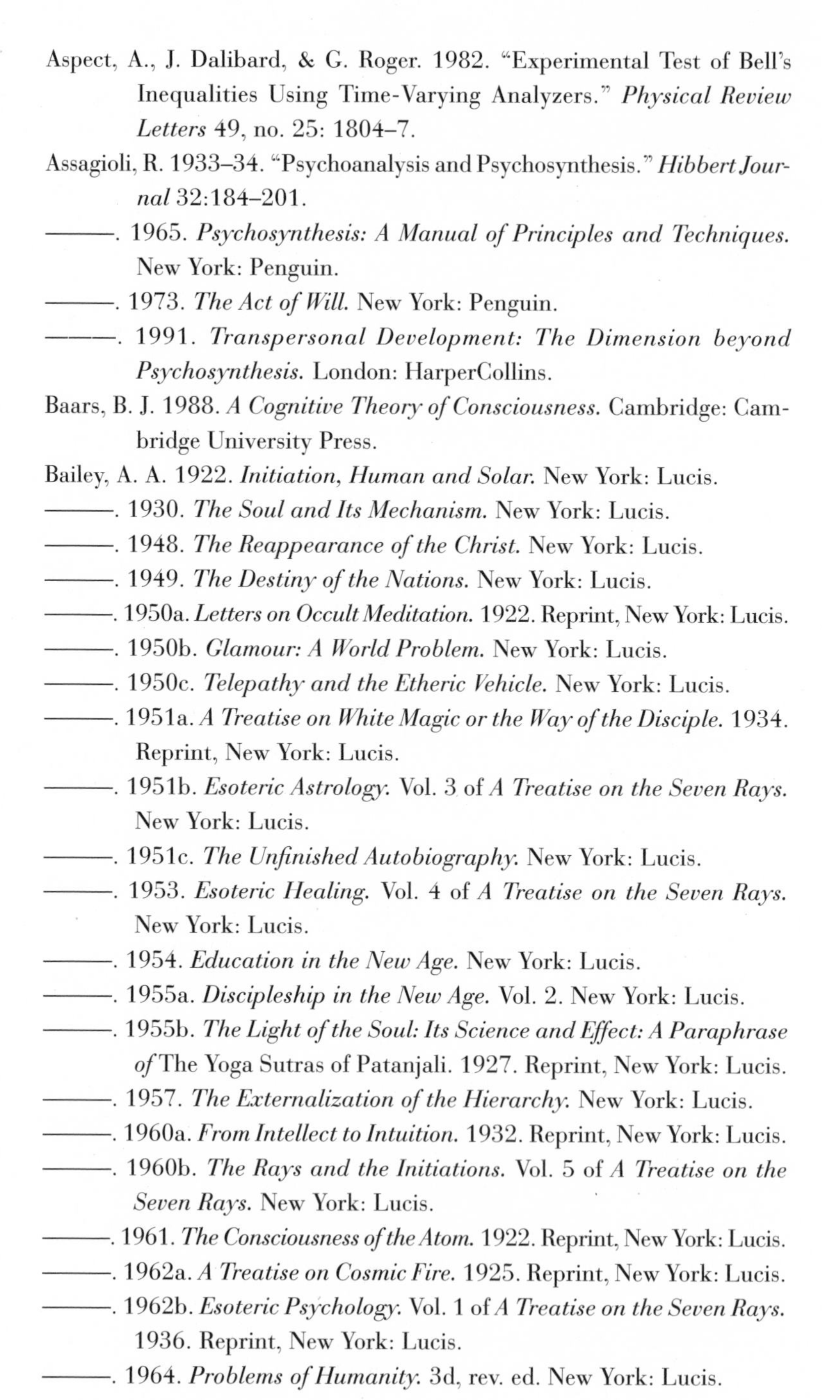

Aspect, A., J. Dalibard, & G. Roger. 1982. "Experimental Test of Bell's Inequalities Using Time-Varying Analyzers." *Physical Review Letters* 49, no. 25: 1804–7.

Assagioli, R. 1933–34. "Psychoanalysis and Psychosynthesis." *Hibbert Journal* 32:184–201.

———. 1965. *Psychosynthesis: A Manual of Principles and Techniques.* New York: Penguin.

———. 1973. *The Act of Will.* New York: Penguin.

———. 1991. *Transpersonal Development: The Dimension beyond Psychosynthesis.* London: HarperCollins.

Baars, B. J. 1988. *A Cognitive Theory of Consciousness.* Cambridge: Cambridge University Press.

Bailey, A. A. 1922. *Initiation, Human and Solar.* New York: Lucis.

———. 1930. *The Soul and Its Mechanism.* New York: Lucis.

———. 1948. *The Reappearance of the Christ.* New York: Lucis.

———. 1949. *The Destiny of the Nations.* New York: Lucis.

———. 1950a. *Letters on Occult Meditation.* 1922. Reprint, New York: Lucis.

———. 1950b. *Glamour: A World Problem.* New York: Lucis.

———. 1950c. *Telepathy and the Etheric Vehicle.* New York: Lucis.

———. 1951a. *A Treatise on White Magic or the Way of the Disciple.* 1934. Reprint, New York: Lucis.

———. 1951b. *Esoteric Astrology.* Vol. 3 of *A Treatise on the Seven Rays.* New York: Lucis.

———. 1951c. *The Unfinished Autobiography.* New York: Lucis.

———. 1953. *Esoteric Healing.* Vol. 4 of *A Treatise on the Seven Rays.* New York: Lucis.

———. 1954. *Education in the New Age.* New York: Lucis.

———. 1955a. *Discipleship in the New Age.* Vol. 2. New York: Lucis.

———. 1955b. *The Light of the Soul: Its Science and Effect: A Paraphrase of* The Yoga Sutras of Patanjali. 1927. Reprint, New York: Lucis.

———. 1957. *The Externalization of the Hierarchy.* New York: Lucis.

———. 1960a. *From Intellect to Intuition.* 1932. Reprint, New York: Lucis.

———. 1960b. *The Rays and the Initiations.* Vol. 5 of *A Treatise on the Seven Rays.* New York: Lucis.

———. 1961. *The Consciousness of the Atom.* 1922. Reprint, New York: Lucis.

———. 1962a. *A Treatise on Cosmic Fire.* 1925. Reprint, New York: Lucis.

———. 1962b. *Esoteric Psychology.* Vol. 1 of *A Treatise on the Seven Rays.* 1936. Reprint, New York: Lucis.

———. 1964. *Problems of Humanity.* 3d, rev. ed. New York: Lucis.

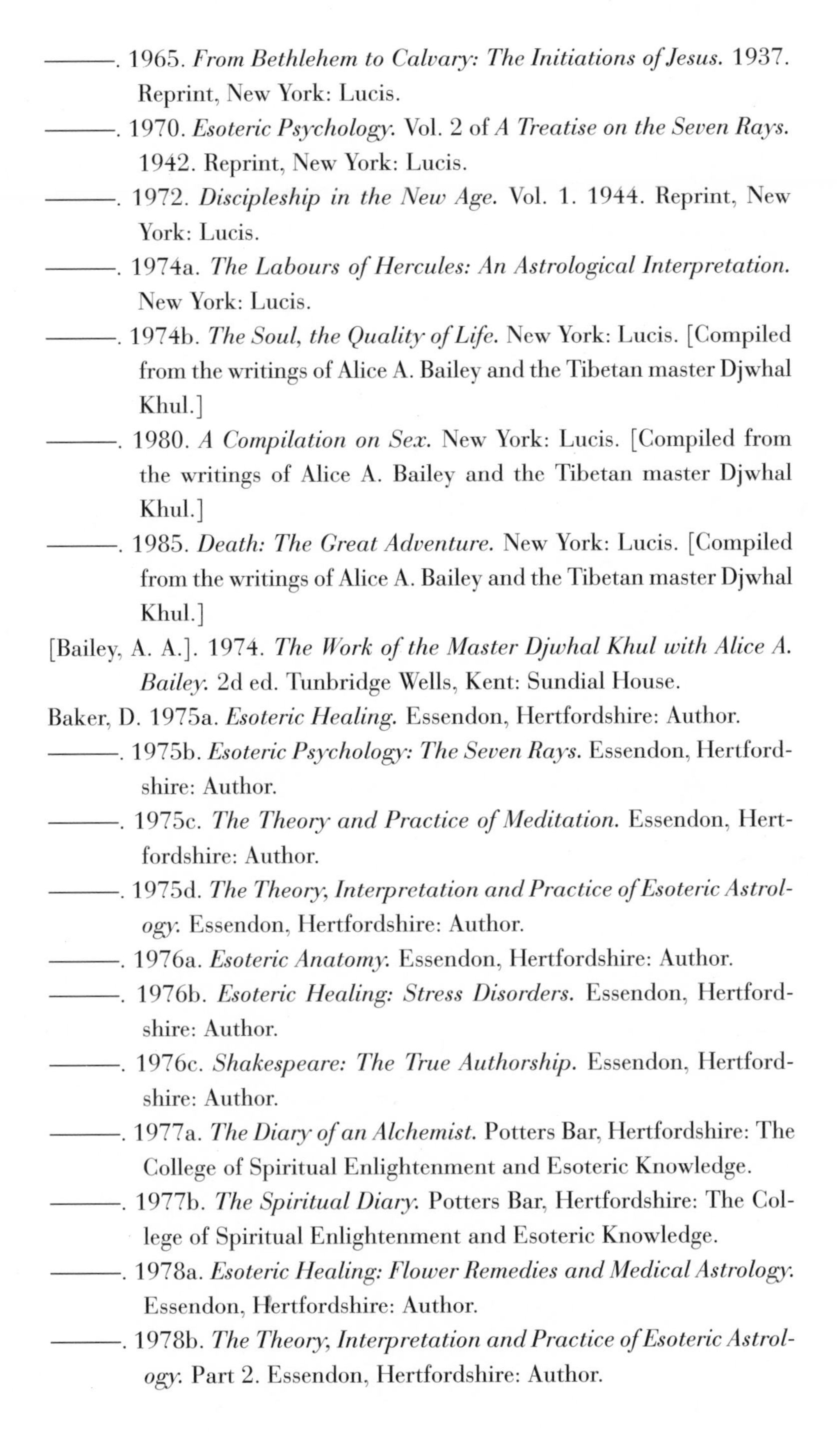

———. 1965. *From Bethlehem to Calvary: The Initiations of Jesus.* 1937. Reprint, New York: Lucis.

———. 1970. *Esoteric Psychology.* Vol. 2 of *A Treatise on the Seven Rays.* 1942. Reprint, New York: Lucis.

———. 1972. *Discipleship in the New Age.* Vol. 1. 1944. Reprint, New York: Lucis.

———. 1974a. *The Labours of Hercules: An Astrological Interpretation.* New York: Lucis.

———. 1974b. *The Soul, the Quality of Life.* New York: Lucis. [Compiled from the writings of Alice A. Bailey and the Tibetan master Djwhal Khul.]

———. 1980. *A Compilation on Sex.* New York: Lucis. [Compiled from the writings of Alice A. Bailey and the Tibetan master Djwhal Khul.]

———. 1985. *Death: The Great Adventure.* New York: Lucis. [Compiled from the writings of Alice A. Bailey and the Tibetan master Djwhal Khul.]

[Bailey, A. A.]. 1974. *The Work of the Master Djwhal Khul with Alice A. Bailey.* 2d ed. Tunbridge Wells, Kent: Sundial House.

Baker, D. 1975a. *Esoteric Healing.* Essendon, Hertfordshire: Author.

———. 1975b. *Esoteric Psychology: The Seven Rays.* Essendon, Hertfordshire: Author.

———. 1975c. *The Theory and Practice of Meditation.* Essendon, Hertfordshire: Author.

———. 1975d. *The Theory, Interpretation and Practice of Esoteric Astrology.* Essendon, Hertfordshire: Author.

———. 1976a. *Esoteric Anatomy.* Essendon, Hertfordshire: Author.

———. 1976b. *Esoteric Healing: Stress Disorders.* Essendon, Hertfordshire: Author.

———. 1976c. *Shakespeare: The True Authorship.* Essendon, Hertfordshire: Author.

———. 1977a. *The Diary of an Alchemist.* Potters Bar, Hertfordshire: The College of Spiritual Enlightenment and Esoteric Knowledge.

———. 1977b. *The Spiritual Diary.* Potters Bar, Hertfordshire: The College of Spiritual Enlightenment and Esoteric Knowledge.

———. 1978a. *Esoteric Healing: Flower Remedies and Medical Astrology.* Essendon, Hertfordshire: Author.

———. 1978b. *The Theory, Interpretation and Practice of Esoteric Astrology.* Part 2. Essendon, Hertfordshire: Author.

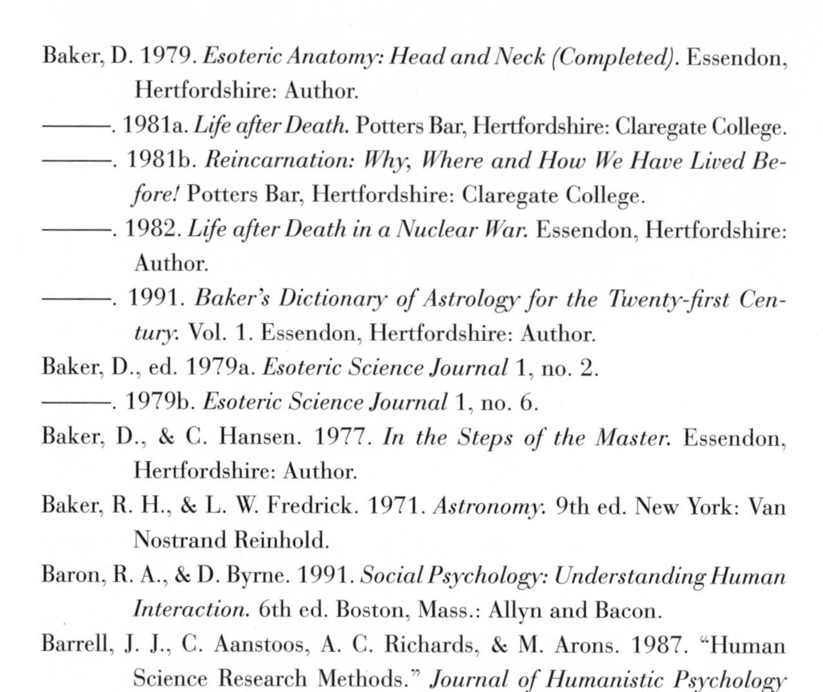

Baker, D. 1979. *Esoteric Anatomy: Head and Neck (Completed).* Essendon, Hertfordshire: Author.

———. 1981a. *Life after Death.* Potters Bar, Hertfordshire: Claregate College.

———. 1981b. *Reincarnation: Why, Where and How We Have Lived Before!* Potters Bar, Hertfordshire: Claregate College.

———. 1982. *Life after Death in a Nuclear War.* Essendon, Hertfordshire: Author.

———. 1991. *Baker's Dictionary of Astrology for the Twenty-first Century.* Vol. 1. Essendon, Hertfordshire: Author.

Baker, D., ed. 1979a. *Esoteric Science Journal* 1, no. 2.

———. 1979b. *Esoteric Science Journal* 1, no. 6.

Baker, D., & C. Hansen. 1977. *In the Steps of the Master.* Essendon, Hertfordshire: Author.

Baker, R. H., & L. W. Fredrick. 1971. *Astronomy.* 9th ed. New York: Van Nostrand Reinhold.

Baron, R. A., & D. Byrne. 1991. *Social Psychology: Understanding Human Interaction.* 6th ed. Boston, Mass.: Allyn and Bacon.

Barrell, J. J., C. Aanstoos, A. C. Richards, & M. Arons. 1987. "Human Science Research Methods." *Journal of Humanistic Psychology* 27, no. 4: 424–57.

Baruss, I. 1990a. *The Personal Nature of Notions of Consciousness: A Theoretical and Empirical Examination of the Role of the Personal in the Understanding of Consciousness.* Lanham, Md.: University Press of America.

———. 1990b. Review of *Consciousness,* by William G. Lycan. *Imagination, Cognition and Personality* 9, no. 2: 179–82.

———. 1992. "Contemporary Issues Concerning the Scientific Study of Consciousness." *Anthropology of Consciousness* 3, nos. 3–4: 28–35.

———. 1993. "Can We Consider Matter as Ultimate Reality? Some Fundamental Problems with a Materialist Interpretation of Reality." *Ultimate Reality and Meaning: Interdisciplinary Studies in the Philosophy of Understanding* 16:245–54.

Baruss, I., & R. J. Moore. 1989. "Notions of Consciousness and Reality." In *Imagery: Current Perspectives,* edited by J. E. Shorr et al., 87–92. New York: Plenum.

———. 1992. "Measurement of Beliefs about Consciousness and Reality." *Psychological Reports* 71:59–64.

Bates, B. C., & A. Stanley. 1985. "The Epidemiology and Differential Diagnosis of Near-Death Experience." *American Journal of Orthopsychiatry* 55, no. 4: 542–49.

Battista, J. R. 1978. "The Science of Consciousness." In *The Stream of Consciousness: Scientific Investigations into the Flow of Human Experience,* edited by K. S. Pope & J. L. Singer, 55–87. New York: Plenum.

Bauer, H. H. 1992. *Scientific Literacy and the Myth of the Scientific Method.* Urbana: University of Illinois Press.

Bauer, M. 1985. "Near-Death Experiences and Attitude Change." *Anabiosis—The Journal for Near-Death Studies* 5, no. 1: 39–47.

Beehler, R. 1990. "Freedom and Authenticity." *Journal of Applied Philosophy* 7, no. 1: 39–44.

Bem, D. J., & C. Honorton. 1994. "Does Psi Exist? Replicable Evidence for an Anomalous Process of Information Transfer." *Psychological Bulletin* 115, no. 1: 4–18.

Benson, D. E., A. Swann, R. O'Toole, & J. P. Turbett. 1991. "Physicians' Recognition of and Response to Child Abuse: Northern Ireland and the U.S.A." *Child Abuse and Neglect* 15:57–67.

Benson, H. 1975. *The Relaxation Response.* New York: William Morrow.

———. 1977. "Transcendental Meditation." *New England Journal of Medicine* 297, no. 9: 513.

———. 1983. "The Relaxation Response: Its Subjective and Objective Historical Precedents and Physiology." *Trends in Neurosciences* 6:281–84.

Benveniste, J. 1993. "Transfer of Biological Activity by Electromagnetic Fields." *Frontier Perspectives* 3, no. 2: 13–15.

Besant, A. 1897. *The Ancient Wisdom: An Outline of Theosophical Teachings.* London: Theosophical Publishing Society.

Bibby, R. W. 1987. *Fragmented Gods: The Poverty and Potential of Religion in Canada.* Toronto: Irwin.

Bickhard, M. H. 1992. "Myths of Science: Misconceptions of Science in Contemporary Psychology." *Theory and Psychology* 2, no. 3: 321–37.

Blackburn, S. 1994. *The Oxford Dictionary of Philosophy.* Oxford: Oxford University Press.

Blackmore, S. J. 1982. *Beyond the Body: An Investigation of Out-of-Body Experiences.* London: Heinemann.

———. 1984. "A Psychological Theory of the Out-of-Body Experience." *Journal of Parapsychology* 48:201–18.

Blanchet, A., A. Dorna, J. Jakobi, I. Lubek, B. Matalon, & H. L. Minton. 1992. "Organizing the International Conference on the History of Social Psychology (Paris, April 18–20, 1991)." *Canadian Psychology/Psychologie canadienne* 33, no. 3: 525–28.

Blanpied, W. A. 1969. *Physics: Its Structure and Evolution.* Waltham, Mass.: Blaisdell.

Blavatsky, H. P. 1962. *The Secret Doctrine: The Synthesis of Science, Religion and Philosophy.* 6 vols. 1888, 1897. Reprint, Adyar, India: Theosophical Publishing House.

———. 1971. *The Theosophical Glossary.* 1892. Reprint, Los Angeles, Calif.: The Theosophy Company.

Bogart, G. C. 1992. "Separating from a Spiritual Teacher." *Journal of Transpersonal Psychology* 24, no. 1: 1–21.

Bohm, D. 1980. *Wholeness and the Implicate Order.* London: Ark.

Boulting, N. E. 1993. "Charles S. Peirce's Idea of Ultimate Reality and Meaning Related to Humanity's Ultimate Future as Seen through Scientific Inquiry." *Ultimate Reality and Meaning* 16, nos. 1–2: 9–26.

Bowers, K. S., G. Regehr, C. Balthazard, & K. Parker. 1990. "Intuition in the Context of Discovery." *Cognitive Psychology* 22:72–110.

Bragdon, E. 1993. *A Sourcebook for Helping People with Spiritual Problems.* Aptos, Calif.: Lightening Up Press.

Braude, S. E. 1988. "Mediumship and Multiple Personality." *Journal of the Society for Psychical Research* 55, no. 813: 177–95.

Bruce, J. 1972. *Travels to Discover the Source of the Nile, in the Years 1768, 1769, 1770, 1771, 1772, and 1773 in five volumes.* 1790. Reprint, Westmead, Farnborough, Hants.: Gregg International.

Byrd, R. C. 1988. "Positive Therapeutic Effects of Intercessory Prayer in a Coronary Care Unit Population." *Southern Medical Journal* 81, no. 7: 826–29.

Caddy, E. 1987. *Opening Doors Within.* Edited and compiled by D. E. Platts. The Park, Forres: Findhorn.

Caird, D. 1987. "Religiosity and Personality: Are Mystics Introverted, Neurotic, or Psychotic?" *British Journal of Social Psychology* 26:345–46.

Campbell, B. F. 1980. *Ancient Wisdom Revived: A History of the Theosophical Movement.* Berkeley and Los Angeles: University of California Press.

Campbell, J., & R. Roberts. 1979. *Tarot Revelations.* San Francisco, Calif.: Alchemy Books.

Carr, C. 1993. "Death and Near-Death: A Comparison of Tibetan and Euro-American Experiences." *Journal of Transpersonal Psychology* 25, no. 1: 59–110.

Carrington, P. 1986. "Meditation as an Access to Altered States of Consciousness." In *Handbook of States of Consciousness,* edited by B. B. Wolman and M. Ullman, 487–523. New York: Van Nostrand Reinhold.

———. 1987. "Managing Meditation in Clinical Practice." In *The Psychology of Meditation,* edited by M. A. West, 150–72. Oxford: Clarendon Press.

Castaneda, C. 1971a. *The Teachings of Don Juan: A Yaqui Way of Knowledge.* Berkeley and Los Angeles: University of California Press.

———. 1971b. *A Separate Reality: Further Conversations with Don Juan.* New York: Simon & Schuster.

———. 1972. *Journey to Ixtlan: The Lessons of Don Juan.* New York: Simon & Schuster.

———. 1974. *Tales of Power.* New York: A Touchstone Book, Simon & Schuster.

Cauthen, N. R., & C. A. Prymak. 1977. "Meditation versus Relaxation: An Examination of the Physiological Effects of Relaxation Training and of Different Levels of Experience with Transcendental Meditation." *Journal of Consulting and Clinical Psychology* 45, no. 3: 496–97.

Chang, S. C. 1978. "The Psychology of Consciousness." *American Journal of Psychotherapy* 32:105–16.

Cheney, C. D. 1993. "Behaviorism: An Overview." In *Survey of Social Science: Psychology Series,* edited by F. N. Magill, 401–7. Pasadena, Calif.: Salem.

Chevalier, G. 1976. *The Sacred Magician: A Ceremonial Diary.* Frogmore, St. Albans, Hertfordshire: Paladin.

Churchill, W. 1992. *Fantasies of the Master Race: Literature, Cinema and the Colonization of American Indians.* Monroe, Maine: Common Courage.

Churchland, P. S. 1980. "A Perspective on Mind-Brain Research." *Journal of Philosophy* 77, no. 4: 185–207.

Cialdini, R. B. 1988. *Influence: Science and Practice.* 2d ed. New York: HarperCollins.

Clark, J. 1992. "Contemplating Nature with Arthur Zajonc." *Noetic Sciences Review* 23:19–28.

Clifton, C. S. 1989. "The Three Faces of Satan: A Close Look at the 'Satanism Scare.'" *Gnosis: A Journal of the Western Inner Traditions,* no. 12:9–18.

Coan, R. W. 1968. "Dimensions of Psychological Theory." *American Psychologist* 23:715–22.

Cohen, A. 1992. *Autobiography of an Awakening.* Corte Madera, Calif.: Moksha Foundation.

Combs, A. L. 1993. "The Collective Unconscious." In *Survey of Social Science: Psychology Series,* edited by F. N. Magill, 592–97. Pasadena, Calif.: Salem.

Combs, A., & M. Holland. 1990. *Synchronicity: Science, Myth, and the Trickster.* New York: Paragon.

Corey, M. A. 1988. "The Psychology of Channeling." *Psychology, A Journal of Human Behavior* 25, nos. 3–4: 86–92.

Corsini, R. J., ed. 1994. *Encyclopedia of Psychology.* 2d ed. New York: John Wiley & Sons.

Cosier, R. A., & J. C. Aplin. 1982. "Intuition and Decision Making: Some Empirical Evidence." *Psychological Reports* 51:275–81.

Creme, B. 1980. *The Reappearance of the Christ and the Masters of Wisdom.* North Hollywood, Calif.: Tara.

Crick, F., & C. Koch. 1992. "The Problem of Consciousness." *Scientific American,* September, 153–59.

Crider, A. B., G. R. Goethals, R. D. Kavanaugh, & P. R. Solomon. 1993. *Psychology.* 4th ed. New York: HarperCollins.

Dane, J. R., & D. E. DeGood. 1987. Review of *Handbook of States of Consciousness,* edited by B. B. Wolman and M. Ullman. *Journal of Parapsychology* 51:91–98.

Davenas, E., F. Beauvais, J. Amara, M. Oberbaum, B. Robinzon, A. Miadonna, A. Tedeschi, B. Pomeranz, P. Fortner, P. Belon, J. Sainte-Laudy, B. Poitevin, & J. Benveniste. 1988. "Human Basophil Degranulation Triggered by Very Dilute Antiserum against IgE." *Nature* 333, no. 6176: 816–18.

Davison, G. C., & J. M. Neale. 1986. *Abnormal Psychology: An Experimental Clinical Approach.* 4th ed. New York: John Wiley.

Dawson, L. L. 1987. "On References to the Transcendent in the Scientific Study of Religion: A Qualified Idealist Proposal." *Religion* 17:227–50.

DeCarvalho, R. J. 1990. "A History of the 'Third Force' in Psychology." *Journal of Humanistic Psychology* 30, no. 4: 22–44.

———. 1991. "'Was Maslow an Aristotelian?' Revisited." *Psychological Record* 41:117–23.

Delaforge, G. 1988. "The Templar Tradition Yesterday and Today." *Gnosis: A Journal of the Western Inner Traditions,* no. 6:8–13.

De Mille, R., ed. 1990. *The Don Juan Papers: Further Castaneda Controveries.* Belmont, Calif.: Wadsworth.

Dennett, D. C. 1978. *Brainstorms: Philosophical Essays on Mind and Psychology.* Montgomery, Vt.: Bradford.

———. 1988. "Quining Qualia." In *Consciousness in Contemporary Science*, edited by A. J. Marcel & E. Bisiach, 42–77. Oxford: Clarendon.

Donner, F. 1991. *Being-in-Dreaming.* San Francisco, Calif.: HarperSanFrancisco.

Dossey, L. 1989. *Recovering the Soul: A Scientific and Spiritual Search.* New York: Bantam.

———. 1991. *Meaning and Medicine: Lessons from a Doctor's Tales of Breakthrough and Healing.* New York: Bantam.

Dowling, L. H. 1964. *The Aquarian Gospel of Jesus the Christ.* 1907. Reprint, Los Angeles, Calif.: DeVorss.

Dunne, B. J., R. G. Jahn, & R. D. Nelson. 1985. *Princeton Engineering Anomalies Research.* Princeton Engineering Anomalies Research Technical Note PEAR 85003. Princeton, N. J.: Princeton University.

Dusen, W. V. 1967. "The Natural Depth in Man." In *Person to Person: The Problem of Being Human: A New Trend in Psychology,* edited by C. R. Rogers & B. Stevens, 211–34. Moab, Utah: Real People.

Einstein, A., B. Podolsky, & N. Rosen. 1983. "Can Quantum-Mechanical Description of Physical Reality Be Considered Complete?" In *Quantum Theory and Measurement,* edited by J. A. Wheeler & W. H. Zurek, 138–41. Princeton, N.J.: Princeton University Press.

Eisenberg, D. 1990. "Energy Medicine in China: Defining a Research Strategy Which Embraces the Criticism of Skeptical Colleagues." *Noetic Sciences Review* 14:4–11.

Eliade, M., ed. 1987. "Soul." In *The Encyclopedia of Religion,* edited by M. Eliade, 13:426–65. New York: Macmillan.

Elitzur, A. C. 1989. "Consciousness and the Incompleteness of the Physical Explanation of Behavior." *Journal of Mind and Behavior* 10, no. 1: 1–20.

Epstein, M. D. 1984. "On the Neglect of Evenly Suspended Attention." *Journal of Transpersonal Psychology* 16, no. 2: 193–205.

Epstein, M. D., & J. D. Lieff. 1981. "Psychiatric Complications of Meditation Practice." *Journal of Transpersonal Psychology* 13, no 2: 137–47.

Ertel, S. 1993. "Comments on Dutch Investigations of the Gauquelin Mars Effect." *Journal of Scientific Exploration* 7, no. 3: 283–92.

Fadiman, J. 1992. Review of *The Future of the Body: Explorations into the Further Evolution of Human Nature,* by Michael Murphy. *Journal of Transpersonal Psychology* 24, no. 2: 212–13.

Fadiman, J. 1993. "Overcoming Abuse." *Association for Humanistic Psychology Perspective*, July/August, 24–25.

Farthing, G. W. 1992. *The Psychology of Consciousness*. Englewood Cliffs, N.J.: Prentice Hall.

Ferrucci, P. 1982. *What We May Be: Techniques for Psychological and Spiritual Growth through Psychosynthesis*. Los Angeles, Calif.: Jeremy P. Tarcher.

———. 1990. *Inevitable Grace: Breakthroughs in the Lives of Great Men and Women: Guides to Your Self-realization*. Translated by D. Kennard. Los Angeles, Calif.: Jeremy P. Tarcher.

Festinger, L., H. W. Riecken, & S. Schachter. 1956. *When Prophecy Fails*. Minneapolis: University of Minnesota Press.

Feynman, R. P., R. B. Leighton, & M. Sands. 1965. *The Feynman Lectures on Physics: Quantum Mechanics*. Reading, Mass.: Addison-Wesley.

Fishman, S. 1990. "The Dean of Psi." *Omni* 12, no. 12 (September): 42–46, 88, 90, 92.

Forman, R. K. C. 1990. "Introduction: Mysticism, Constructivism, and Forgetting." In *The Problem of Pure Consciousness: Mysticism and Philosophy*, edited by R. K. C. Forman, 3–49. New York: Oxford University Press.

Foundation for Inner Peace. 1975. *A Course in Miracles*. Farmingdale, N.Y.: Author.

Fox, O. 1962. *Astral Projection: A Record of Out-of-the-Body Experiences*. Secaucus, N.J.: Citadel.

Frank, J. D. 1977. "Nature and Functions of Belief Systems: Humanism and Transcendental Religion." *American Psychologist* 32, no. 7: 555–59.

Frankl, V. E. 1966. "Self-transcendence as a Human Phenomenon." *Journal of Humanistic Psychology* 6, no. 2: 97–106.

———. 1984. *Man's Search for Meaning*. New York: Simon & Schuster.

Fuller, J. G. 1985. *The Ghost of Twenty-nine Megacycles: A New Breakthrough in Life after Death?* London: Souvenir.

Geller, L. 1982. "The Failure of Self-actualization Theory: A Critique of Carl Rogers and Abraham Maslow." *Journal of Humanistic Psychology* 22, no. 2: 56–73.

Gescheider, G. A. 1985. *Psychophysics: Method, Theory, and Application*. 2d ed. Hillsdale, N.J.: Lawrence Erlbaum.

Giorgi, A. P. 1994. "Existentialism." In *Encyclopedia of Psychology*, edited by R. J. Corsini, 1:520–21. 2d ed. New York: John Wiley & Sons.

Giovacchini, P. L. 1987. *A Narrative Textbook of Psychoanalysis.* Northvale, N.J.: Jason Aronson.

Goleman, D. 1987. "Carl R. Rogers, 85, Leader in Psychotherapy, Dies." *New York Times Biographical Service* 18, no. 2: 96–97.

———. 1988. *The Meditative Mind: The Varieties of Meditative Experience.* Los Angeles, Calif.: Jeremy P. Tarcher.

Gomes, M. 1987. *The Dawning of the Theosophical Movement.* Wheaton, Ill.: Theosophical Publishing House, Quest.

Goring, R., ed. 1994. *Larousse Dictionary of Beliefs and Religions.* New York: Larousse.

Goudge, T. A. 1967. "Teilhard de Chardin, Pierre." In *The Encyclopedia of Philosophy,* edited by P. Edwards, 8:83–84. New York: Macmillan; Free Press.

Graves, F. D. 1973. *The Windows of Tarot.* Dobbs Ferry, N.Y.: Morgan & Morgan.

Greeley, A. 1987. "The 'Impossible': It's Happening." *Noetic Sciences Review* 2:7–9.

Green, E. E., & A. M. Green. 1986. "Biofeedback and States of Consciousness." In *Handbook of States of Consciousness,* edited by B. B. Wolman & M. Ullman, 553–89. New York: Van Nostrand Reinhold.

Green, M. B., J. H. Schwarz, & E. Witten. 1987. *Superstring Theory.* Cambridge: Cambridge University Press.

Greenberg, D., E. Witztum, & J. Pisante. 1987. "Scrupulosity: Religious Attitudes and Clinical Presentations." *British Journal of Medical Psychology* 60, no. 1: 29–37.

Greening, T. C. 1975. "In Memorium: Roberto Assagioli." *Journal of Humanistic Psychology* 15, no. 1: 4–5.

Guiley, R. E. 1991. *Harper's Encyclopedia of Mystical and Paranormal Experience.* San Franciso, Calif.: HarperSanFrancisco.

Haden, N. K. 1989. "Of Paradigms, Saints, and Individuals: The Question of Authenticity." *Dialogue: Journal of Phi Sigma Tau* 32, no. 1: 7–14.

Haisch, B. 1993. "Society for Scientific Exploration: Position Paper." *Journal of Scientific Exploration* 7, no. 1: 107.

Hanson, N. R. 1963. "The Dematerialization of Matter." In *The Concept of Matter,* edited by E. McMullin, 549–61. Notre Dame, Ind.: University of Notre Dame Press.

Hanson, S. J., & D. J. Burr. 1990. "What Connectionist Models Learn: Learning and Representation in Connectionist Networks." *Behavioral and Brain Sciences* 13, no. 3: 471–89.

Harary, K., & P. Weintraub. 1989. *Have an Out-of-Body Experience in Thirty Days: The Free Flight Program.* New York: St. Martin's.

Hardy, J. 1987. *A Psychology with a Soul: Psychosynthesis in Evolutionary Context.* London: Arkana.

Harman, W. W. 1987. "Survival of Consciousness after Death: A Perennial Issue Revisited." In *Consciousness and Survival: An Interdisciplinary Inquiry into the Possibility of Life beyond Biological Death,* edited by J. S. Spong, 1–10. Sausalito, Calif.: Institute of Noetic Sciences.

Harman, W., & H. Rheingold. 1984. *Higher Creativity: Liberating the Unconscious for Breakthrough Insights.* Los Angeles, Calif.: Jeremy P. Tarcher.

Harsch-Fischbach, M. 1989. *Cercle d'Etudes sur la Transcommunication—Luxembourg Newsletter Issue* 01/89. Translated by H. Heckmann.

———. 1992. *Cercle d'Etudes sur la Transcommunication—Luxembourg INFOnews Issue* 02/92. Translated by H. Heckmann.

Heckler, R. S. 1990. *In Search of the Warrior Spirit.* Berkeley, Calif.: North Atlantic.

Heidegger, M. 1962. *Being and Time.* Translated by J. Macquarrie & E. Robinson. New York: Harper & Row.

Helminiak, D. A. 1984. "Consciousness as a Subject Matter." *Journal for the Theory of Social Behaviour* 14, no. 2: 211–30.

———. 1987. *Spiritual Development: An Interdisciplinary Study.* Chicago, Ill.: Loyola University Press.

Hieratic. 1975. *The Complete Planetary Ephemeris for 1950 to 2000 A.D.* Medford, Mass.: Author.

Hilgard, E. R. 1980. "Consciousness in Contemporary Psychology." *Annual Review of Psychology* 31:1–26.

Hill, O. W., Jr. 1987. "Intuition: Inferential Heuristic or Epistemic Mode?" *Imagination, Cognition and Personality* 7, no. 2: 137–54.

Hofstadter, D. R. 1979. *Gödel, Escher, Bach: An Eternal Golden Braid.* New York: Basic.

Holmes, D. S. 1987. "The Influence of Meditation versus Rest on Physiological Arousal: A Second Examination." In *The Psychology of Meditation,* edited by M. A. West, 81–103. Oxford: Clarendon Press.

Hughes, D. J. 1991. "Blending with an Other: An Analysis of Trance Channeling in the United States." *Ethos: Journal of the Society for Psychological Anthropology* 19, no. 2: 161–84.

Hutch, R. A. 1980. "Helena Blavatsky Unveiled." *Journal of Religious History* 11, no. 2: 320–41.

Huxley, A. 1946. *The Perennial Philosophy*. London: Chatto & Windus.

Ingram, C. 1992. "Plunge into Eternity: An Interview with H. L. Poonja by Catherine Ingram." *Yoga Journal* 106 (September/October): 56–63.

Jahn, R. G., & B. J. Dunne. 1987. *Margins of Reality: The Role of Consciousness in the Physical World.* San Diego, Calif.: Harcourt Brace Jovanovich.

Jahn, R. G., B. J. Dunne, & R. D. Nelson. 1987. "Engineering Anomalies Research." *Journal of Scientific Exploration* 1, no. 1: 21–50.

James, W. 1904. "Does 'Consciousness' Exist?" *Journal of Philosophy, Psychology and Scientific Methods* 1, no. 18: 477–91.

———. 1958. *The Varieties of Religious Experience: A Study in Human Nature. Being the Gifford Lectures on Natural Religion Delivered at Edinburgh in 1901–1902.* 1902. Reprint, New York: NAL Penguin.

———. 1983. *The Principles of Psychology*. 1890. Reprint, Cambridge, Mass.: Harvard University Press.

Johnson, G. 1990. "Telepathic Dreams." *Consciousness Review* 1:37–46.

Josephson, B. D., & B. A. Rubik. 1992. "The Challenge of Consciousness Research." *Frontier Perspectives* 3, no. 1: 15–19.

Joy, W. B. 1985. "The New Age, Armageddon, and Mythic Cycles." *Journal of Humanistic Psychology* 25, no. 1: 85–89.

Kafatos, M., & R. Nadeau. 1990. *The Conscious Universe: Part and Whole in Modern Physical Theory*. New York: Springer-Verlag.

Keen, E. 1975. *A Primer in Phenomenological Psychology*. Lanham, Md.: University Press of America.

Keutzer, C. S. 1984. "The Power of Meaning: From Quantum Mechanics to Synchronicity." *Journal of Humanistic Psychology* 24, no. 1: 80–94.

Kimble, G. A. 1984. "Psychology's Two Cultures." *American Psychologist* 39, no. 8: 833–39.

Kinney, J. 1988. "Human Rites & Hidden Assets." *Gnosis: A Journal of the Western Inner Traditions*, no. 6:6–7.

Kinney, J., & T. O'Neill. 1989. "The Imperator of AMORC: An Interview with Gary L. Stewart." *Gnosis: A Journal of the Western Inner Traditions*, no. 12:33–35.

Kirschenbaum, H., & V. L. Henderson, eds. 1989. *The Carl Rogers Reader*. Boston, Mass.: Houghton Mifflin.

Klein, B. 1984. *Movements of Magic: The Spirit of T'ai-chi-Ch'uan.* North Hollywood, Calif.: Newcastle.

Klein, D. B. 1984. *The Concept of Consciousness: A Survey*. Lincoln: University of Nebraska Press.

Klimo, J. 1987. *Channeling: Investigations on Receiving Information from Paranormal Sources.* Los Angeles, Calif.: Jeremy P. Tarcher.

Kornfield, J. 1993. *A Path with Heart: A Guide through the Perils and Promises of Spiritual Life.* New York: Bantam.

Kramer, J., & D. Alstad. 1993. *The Guru Papers: Masks of Authoritarian Power.* Berkeley, Calif.: Frog.

Krasner, L., & A. C. Houts. 1984. "A Study of the 'Value' Systems of Behavioral Scientists." *American Psychologist* 39, no 8: 840–50.

Kremer, J. W. 1992. "Lifework: Carlos Castaneda." *ReVision: A Journal of Consciousness and Transformation* 14, no. 4: 195–203.

Krieger, D. 1981. *Foundations for Holistic Health Nursing Practices: The Renaissance Nurse.* Philadelphia, Pa.: J. B. Lippincott.

Krippner, S., & L. George. 1986. "Psi Phenomena as Related to Altered States of Consciousness." In *Handbook of States of Consciousness,* edited by B. B. Wolman & M. Ullman, 332–64. New York: Van Nostrand Reinhold.

Kuhn, T. S. 1970. *The Structure of Scientific Revolutions.* 2d ed. Chicago, Ill.: University of Chicago Press.

Kukla, A. 1983. "Toward a Science of Experience." *Journal of Mind and Behavior* 4, no. 2: 231–46.

LaBerge, S., & J. Gackenbach. 1986. "Lucid Dreaming." In *Handbook of States of Consciousness,* edited by B. B. Wolman & M. Ullman, 159–98. New York: Van Nostrand Reinhold.

Lacombe, M. 1982. "Theosophy and the Canadian Idealist Tradition: A Preliminary Exploration." *Journal of Canadian Studies/Revue d'études canadiennes* 17, no. 2: 100–118.

Latané, B., & J. M. Darley. 1968. "Group Inhibition of Bystander Intervention in Emergencies." *Journal of Personality and Social Psychology* 10, no. 3: 215–21.

Leadbeater, C. W. 1915. *The Astral Plane: Its Scenery, Inhabitants, and Phenomena.* London: Theosophical Publishing Society.

———. 1988. *The Masters and the Path (An Abridgement).* 1925. Reprint, Adyar, India: Theosophical Publishing House.

Leighton, A. H. 1990. "Contributions of Epidemiology to Psychiatric Thought." *Canadian Journal of Psychiatry* 35:385–89.

Leonard, R. 1987. "Appendix C: Genesis and Symbolism of Wolff's Mandala." Unpublished paper.

———. 1991. "The Transcendental Philosophy of Franklin Merrell-Wolff." Ph.D. diss., University of Waterloo, Waterloo, Ontario.

Lester, D., J. S. Thinschmidt, & L. A. Trautman. 1987. "Paranormal Belief and Jungian Dimensions of Personality." *Psychological Reports* 61:182.

Levitan, L., & S. LaBerge. 1991. "Mind in Body or Body in Mind? OBEs and Lucid Dreams, Part 2." *Nightlight: The Lucidity Institute Newsletter* 3, no. 3: 1–3.

Lewin, R. 1980. "Is Your Brain Really Necessary?" *Science* 210, no. 12: 1232–34.

Lilly, J. C. 1978. *The Scientist: A Novel Autobiography*. Philadelphia, Pa.: J. B. Lippincott.

Lindauer, M. S. 1994. "Phenomenological Method." In *Encyclopedia of Psychology*, edited by R. J. Corsini, 3:69–70. 2d ed. New York: John Wiley & Sons.

Logan, B. A. 1993. "The 17 KeV Neutrino Controversy." *Physics in Canada/La physique au Canada* 49, no. 1: 21–24.

Lorber, J. 1965. "Hydranencephaly with Normal Development." *Developmental Medicine and Child Neurology* 7:628–33.

———. 1968. "The Results of Early Treatment of Extreme Hydrocephalus." *Developmental Medical Child Neurology Supplement* 16:21–29.

Lorber, J., & R. B. Zachary. 1968. "Primary Congenital Hydrocephalus: Long-Term Results of Controlled Therapeutic Trial." *Archives of Disease in Childhood* 43:516–27.

Lycan, W. G. 1987. *Consciousness.* Cambridge, Mass.: MIT Press, Bradford.

Lyons, W. 1986. *The Disappearance of Introspection.* Cambridge, Mass.: MIT Press, Bradford.

Macy, M. 1993. "When Dimensions Cross." *Noetic Sciences Review* 25:17–20.

Mahadevan, T. M. P. 1977. *Ramaṇa Maharshi: The Sage of Aruṇācala.* London: Allen & Unwin, Mandala Books.

Maher, B. A. 1979. "The Shattered Language of Schizophrenia." In *Consciousness: Brain, States of Awareness, and Mysticism,* edited by D. Goleman & R. J. Davidson, 115–19. New York: Harper & Row.

Malamud, J. R. 1986. "Becoming Lucid in Dreams and Waking Life." In *Handbook of States of Consciousness,* edited by B. B. Wolman & M. Ullman, 590–612. New York: Van Nostrand Reinhold.

Mandell, A. J. 1980. "Toward a Psychobiology of Transcendence: God in the Brain." In *The Psychobiology of Consciousness,* edited by J. M. Davidson & R. J. Davidson, 379–464. New York: Plenum.

Mandler, G. 1985. *Cognitive Psychology: An Essay in Cognitive Science.* Hillsdale, N.J.: Lawrence Erlbaum.

Mann, J. H. 1994. "Yoga." In *Encyclopedia of Psychology*, edited by R. J. Corsini, 3:591–92. 2d ed. New York: John Wiley & Sons.

Mansbach, A. 1991. "Heidegger on the Self, Authenticity and Inauthenticity." *Iyyun, The Jerusalem Philosophical Quarterly* 40:65–91.

Maslow, A. H. 1964. *Religions, Values, and Peak-Experiences*. New York: Penguin.

———. 1966a. "Comments on Dr. Frankl's Paper." *Journal of Humanistic Psychology* 6, no. 2: 107–12.

———. 1966b. *The Psychology of Science: A Reconnaissance*. New York: Harper & Row.

———. 1968. *Toward a Psychology of Being*. 2d ed. New York: Van Nostrand Reinhold.

———. 1971. *The Farther Reaches of Human Nature*. New York: Penguin.

Massaro, D. W., & N. Cowan. 1993. "Information Processing Models: Microscopes of the Mind." *Annual Review of Psychology* 44:383–425.

Maxwell, M., & V. Tschudin. 1990. *Seeing the Invisible: Modern Religious and Other Transcendent Experiences*. London: Penguin, Arkana.

McIntosh, C. 1988. "The Rosicrucian Dream." *Gnosis: A Journal of the Western Inner Traditions*, no. 6:14–19.

———. 1989. "The Modern Rosicrucians." *Gnosis: A Journal of the Western Inner Traditions*, no. 12:26–32.

Melton, J. G., J. Clark, & A. A. Kelly. 1991. *New Age Almanac*. New York: Visible Ink.

Merrell-Wolff, F. 1917a. "Special Courses of Instruction: Occult Mathematics, Series One, Lesson One, Introduction." Unpublished manuscript.

———. 1917b. "Special Courses of Instruction: Occult Mathematics, Series One, Lesson Two, Mathematical Subject-Matter." Unpublished manuscript.

———. 1917c. "Special Courses of Instruction: Occult Mathematics, Series One, Lesson Three, Interpretation of Direct Operations." Unpublished manuscript.

———. 1917d. "Special Courses of Instruction: Occult Mathematics, Series One, Lesson Four, Interpretation of the Numerical Elements." Unpublished manuscript.

———. 1917e. "Special Courses of Instruction: Occult Mathematics, Series One, Lesson Five, Interpretation of the Numerical Elements, Number One." Unpublished manuscript.

———. 1917f. "Special Courses of Instruction: Occult Mathematics, Series One, Lesson Six, Interpretation of the Numerical Elements, Number Two." Unpublished manuscript.

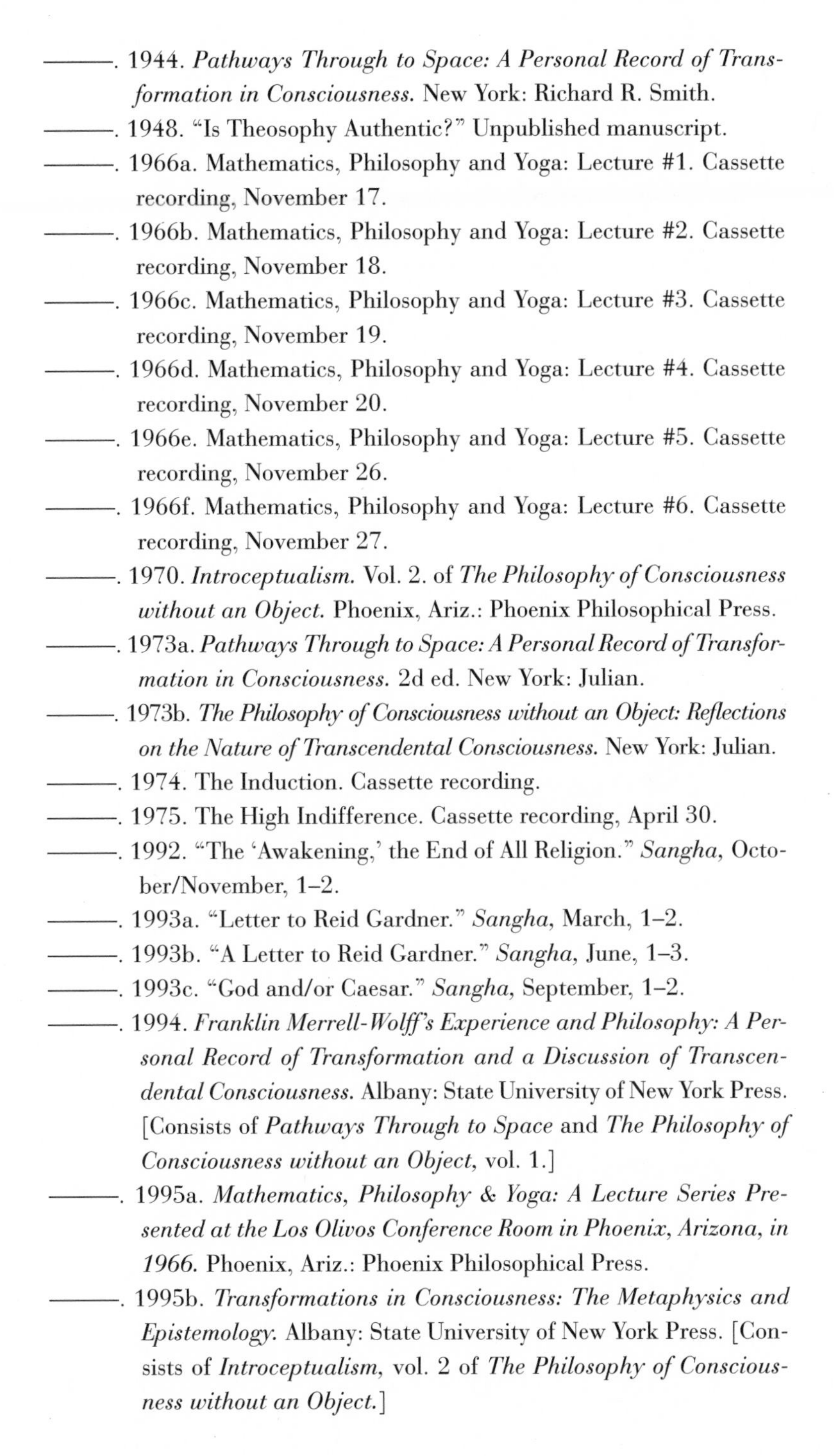

———. 1944. *Pathways Through to Space: A Personal Record of Transformation in Consciousness.* New York: Richard R. Smith.

———. 1948. "Is Theosophy Authentic?" Unpublished manuscript.

———. 1966a. Mathematics, Philosophy and Yoga: Lecture #1. Cassette recording, November 17.

———. 1966b. Mathematics, Philosophy and Yoga: Lecture #2. Cassette recording, November 18.

———. 1966c. Mathematics, Philosophy and Yoga: Lecture #3. Cassette recording, November 19.

———. 1966d. Mathematics, Philosophy and Yoga: Lecture #4. Cassette recording, November 20.

———. 1966e. Mathematics, Philosophy and Yoga: Lecture #5. Cassette recording, November 26.

———. 1966f. Mathematics, Philosophy and Yoga: Lecture #6. Cassette recording, November 27.

———. 1970. *Introceptualism.* Vol. 2. of *The Philosophy of Consciousness without an Object.* Phoenix, Ariz.: Phoenix Philosophical Press.

———. 1973a. *Pathways Through to Space: A Personal Record of Transformation in Consciousness.* 2d ed. New York: Julian.

———. 1973b. *The Philosophy of Consciousness without an Object: Reflections on the Nature of Transcendental Consciousness.* New York: Julian.

———. 1974. The Induction. Cassette recording.

———. 1975. The High Indifference. Cassette recording, April 30.

———. 1992. "The 'Awakening,' the End of All Religion." *Sangha,* October/November, 1–2.

———. 1993a. "Letter to Reid Gardner." *Sangha,* March, 1–2.

———. 1993b. "A Letter to Reid Gardner." *Sangha,* June, 1–3.

———. 1993c. "God and/or Caesar." *Sangha,* September, 1–2.

———. 1994. *Franklin Merrell-Wolff's Experience and Philosophy: A Personal Record of Transformation and a Discussion of Transcendental Consciousness.* Albany: State University of New York Press. [Consists of *Pathways Through to Space* and *The Philosophy of Consciousness without an Object,* vol. 1.]

———. 1995a. *Mathematics, Philosophy & Yoga: A Lecture Series Presented at the Los Olivos Conference Room in Phoenix, Arizona, in 1966.* Phoenix, Ariz.: Phoenix Philosophical Press.

———. 1995b. *Transformations in Consciousness: The Metaphysics and Epistemology.* Albany: State University of New York Press. [Consists of *Introceptualism,* vol. 2 of *The Philosophy of Consciousness without an Object.*]

Messer, S. B. 1985. "Choice of Method Is Value Laden Too." *American Psychologist* 40, no. 12: 1414–15.

Metz, M. S. 1974. *Ephemeriden, 1890–1950.* Zurich: Verlag Max S. Metz.

Michaels, R. R., M. J. Huber, & D. S. McCann. 1976. "Evaluation of Transcendental Meditation as a Method of Reducing Stress." *Science* 192:1242–44.

Miller, R. N. 1982. "Study of the Effectiveness of Remote Mental Healing." *Medical Hypotheses* 8:481–90.

Ming-Dao, D. 1983. *The Wandering Taoist.* New York: Harper & Row.

Minton, H. L. 1992. "Root Metaphors and the Evolution of American Social Psychology." *Canadian Psychology/Psychologie canadienne* 33, no. 3: 547–53.

Mishlove, J. 1993. *The Roots of Consciousness: The Classic Encyclopedia of Consciousness Studies.* Revised and expanded ed. Tulsa, Okla.: Council Oak.

Mitchell, J. L. 1981. *Out-of-Body Experiences: A Handbook.* New York: Ballantine.

Monroe, R. A. 1972. *Journeys Out of the Body.* London: Transworld, Corgi Books.

Moody, R. A., Jr. 1988. *The Light Beyond.* New York: Bantam.

Moody, R. A., Jr., & J. Mishlove. 1988. *Life after Life: Understanding Near-death Experience.* Oakland, Calif.: Innerwork. Videocassette.

Moritz, C., ed. 1962. "Rogers, Carl R(ansom)." *Current Biography* 23, no. 11: 11–13.

Morris, S. 1984. "Games." *Omni* 6, no. 4 (January): 82, 128–29.

Morse, M. 1990. *Closer to the Light: Learning from the Near-Death Experiences of Children.* New York: Ballantine, Ivy Books.

———. 1992. *Transformed by the Light: The Powerful Effect of Near-Death Experiences on People's Lives.* New York: Villard.

Motoyama, H. 1978. *Science and the Evolution of Consciousness: Chakras, Ki, and Psi.* Brookline, Mass.: Autumn.

Murphy, M. 1992a. *The Future of the Body: Explorations into the Further Evolution of Human Nature.* New York: Jeremy P. Tarcher/Perigree.

———. 1992b. "The Future of the Body: Michael Murphy in Conversation with Alexander Blair-Ewart." *Dimensions* 7, no. 10: 32–38.

Murphy, M., & S. Donovan. 1983. "A Bibliography of Meditation Theory and Research: 1931–1983." *Journal of Transpersonal Psychology* 15, no. 2: 181–228.

———. 1988. *The Physical and Psychological Effects of Meditation: A Review of Contemporary Meditation Research with a Comprehensive Bibliography, 1931–1988.* San Rafael, Calif.: Esalen Institute.

Myers, N., ed. 1984. *Gaia: An Atlas of Planet Management.* Garden City, N.Y.: Anchor Press/Doubleday, Anchor Books.

Myers, S. S., & H. Benson. 1992. "Psychological Factors in Healing: A New Perspective on an Old Debate." *Behavioral Medicine* 18:5–11.

Natsoulas, T. 1978a. "Consciousness." *American Psychologist* 33:906–14.

———. 1978b. "Residual Subjectivity." *American Psychologist* 33:269–83.

———. 1986. "Consciousness: Consideration of a Self-intimational Hypothesis." *Journal for the Theory of Social Behaviour* 16, no. 2: 197–207.

———. 1992. "'I Am Not the Subject of This Thought': Understanding a Unique Relation of Special Ownership with the Help of David Woodruff Smith: Part 1." *Imagination, Cognition and Personality* 11, no. 3: 279–302.

Needleman, J. 1965. *A Sense of the Cosmos: The Encounter of Modern Science and Ancient Truth.* New York: E. P. Dutton.

Nelson, P. L. 1989. "Personality Factors in the Frequency of Reported Spontaneous Praeternatural Experiences." *Journal of Transpersonal Psychology* 21, no. 2: 193–209.

———. 1990. "The Technology of the Praeternatural: An Empirically Based Model of Transpersonal Experiences." *Journal of Transpersonal Psychology* 22, no. 1: 35–50.

Nelson, R. D., Y. H. Dobyns, B. J. Dunne, & R. G. Jahn. 1991. *Analysis of Variance of REG Experiments: Operator Intention, Secondary Parameters Database Structure.* Princeton Engineering Anomalies Research Technical Note PEAR 91004. Princeton, N.J.: Princeton University.

Nienhuys, J. W. 1993. "Dutch Investigations of the Gauquelin Mars Effect." *Journal of Scientific Exploration* 7, no. 3: 271–81.

Northrop, F. S. C. 1966. *The Meeting of East and West: An Inquiry Concerning World Understanding.* 1946. Reprint, New York: Collier.

Novak, P. 1987. "Attention." In *The Encyclopedia of Religion,* edited by M. Eliade, 1:501–9. New York: Macmillan.

Olson, J. M., & M. P. Zanna. 1993. "Attitudes and Attitude Change." *Annual Review of Psychology* 44:117–54.

O'Regan, B. 1988. "*Nature* vs. Nature: Science, Censorship and New Ideas." *Noetic Sciences Review* 8:10–13, 26–29.

O'Regan, B., ed. 1985. "Multiple Personality—Mirrors of a New Model of Mind?" *Investigations* 1, nos. 3–4: 1–23.

Ornstein, R. E. 1972. *The Psychology of Consciousness.* New York: Viking.

Osborne, J. 1981. "Approaches to Consciousness in North American Academic Psychology." *Journal of Mind and Behavior* 2, no. 3: 271–91.

Ossoff, J. 1993. "Reflections of Shaktipat: Psychosis or the Rise of Kundalini? A Case Study." *Journal of Transpersonal Psychology* 25, no. 1: 29–42.

Pallak, M. S., D. A. Cook, & J. J. Sullivan. 1980. "Commitment and Energy Conservation." *Applied Social Psychology Annual* 1:235–53.

Pasricha, S. 1992. "Are Reincarnation Type Cases Shaped by Parental Guidance? An Empirical Study Concerning the Limits of Parents' Influence on Children." *Journal of Scientific Exploration* 6, no. 2: 167–80.

Peat, F. D. 1987. *Synchronicity: The Bridge between Matter and Mind.* New York: Bantam.

Pekala, R. J. 1991. *Quantifying Consciousness: An Empirical Approach.* New York: Plenum.

Pekala, R. J., & R. L. Levine. 1981. "Mapping Consciousness: Development of an Empirical-Phenomenological Approach." *Imagination, Cognition and Personality* 1, no. 1: 29–47.

Pelletier, K. R. 1985. *Toward a Science of Consciousness.* 1978. Reprint, Berkeley, Calif.: Celestial Arts.

Perry, M. 1990. "Possession?" *Parapsychology Review* 21, no. 2: 1–4.

Persinger, M. A. 1992. "Neuropsychological Profiles of Adults Who Report 'Sudden Remembering' of Early Childhood Memories: Implications for Claims of Sex Abuse and Alien Visitation/Abduction Experiences." *Perceptual and Motor Skills* 75:259–66.

Phares, E. J. 1994. "Locus of Control." In *Encyclopedia of Psychology,* edited by R. J. Corsini, 2:347–49. 2d ed. New York: John Wiley & Sons.

Phillips, S. M. 1980. *Extra-sensory Perception of Quarks.* Madras, India: Theosophical Publishing House.

Picknett, L. 1990. *The Encyclopaedia of the Paranormal: A Complete Guide to the Unexplained.* London: Macmillan London.

Poloma, M. M., & B. F. Pendleton. 1991. "The Effects of Prayer and Prayer Experiences on Measures of General Well-Being." *Journal of Psychology and Theology* 19, no. 1: 71–83.

Raimy, V. 1994. "Autogenic Training." In *Encyclopedia of Psychology,* edited by R. J. Corsini, 1:124–25. 2d ed. New York: John Wiley & Sons.

Rayburn, C. A. 1993. Review of *Spiritual Development: An Interdisciplinary Study*, by Daniel A. Helminiak. *International Journal for the Psychology of Religion* 3, no. 4: 263–65.

Reason, P., & J. Rowan, eds. 1981. *Human Inquiry: A Sourcebook of New Paradigm Research.* Chichester: John Wiley & Sons.

Regis, E. 1987. *Who Got Einstein's Office? Eccentricity and Genius at the Institute for Advanced Study.* New York: Addison-Wesley.

Richeport, M. M. 1992. "The Interface between Multiple Personality, Spirit Mediumship, and Hypnosis." *American Journal of Clinical Hypnosis* 34, no. 3: 168–77.

Ring, K. 1987. "Near-Death Experiences: Intimations of Immortality?" In *Consciousness and Survival: An Interdisciplinary Inquiry into the Possibility of Life beyond Biological Death*, edited by J. S. Spong, 165–76. Sausalito, Calif.: Institute of Noetic Sciences.

———. 1993. "Near-Death, UFO Events Leave Legacy of Change." *Brain/Mind and Common Sense* 18, no. 6: 3.

Roberts, J. 1978. *The Afterdeath Journal of an American Philosopher: The World View of William James.* Englewood Cliffs, N.J.: Prentice-Hall.

Robson, V. E. 1976. *A Beginner's Guide to Practical Astrology.* New York: Samuel Weiser.

Rodin, J., & P. Salovey. 1989. "Health Psychology." *Annual Review of Psychology* 40:533–579.

Rogers, C. R. 1961. *On Becoming a Person: A Therapist's View of Psychotherapy.* Boston, Mass.: Houghton Mifflin.

———. 1967a. Foreword to *Person to Person: The Problem of Being Human: A New Trend in Psychology*, by C. R. Rogers & B. Stevens. Moab, Utah: Real People.

———. 1967b. "Toward a Modern Approach to Values: The Valuing Process in the Mature Person." In *Person to Person: The Problem of Being Human: A New Trend in Psychology*, by C. R. Rogers & B. Stevens, 13–28. Moab, Utah: Real People.

———. 1980. *A Way of Being.* Boston, Mass.: Houghton Mifflin.

Rogers, C. R., & B. Stevens. 1967. *Person to Person: The Problem of Being Human: A New Trend in Psychology.* Moab, Utah: Real People.

Rosen, J. N. 1953. *Direct Analysis: Selected Papers.* New York: Grune & Stratton.

Rosenfeld, L. 1983. "Bohr's Reply." In *Quantum Theory and Measurement*, edited by J. A. Wheeler & W. H. Zurek, 142–43. Princeton, N.J.: Princeton University Press.

Rowe, J. 1974. "A Mathematical Supplement to the Lectures of Franklin Merrell-Wolff." Unpublished manuscript.

Russell, P. 1976. *The TM Technique: An Introduction to Transcendental Meditation and the Teachings of Maharishi Mahesh Yogi.* London: Routledge & Kegan Paul.

Ryder, D. 1992. *Breaking the Circle of Satanic Ritual Abuse: Recognizing and Recovering from the Hidden Trauma.* Minneapolis, Minn.: Compcare.

Sannella, L. 1976. *Kundalini—Psychosis or Transcendence?* San Francisco, Calif.: H. S. Dakin.

Schneider, J., C. W. Smith, C. Minning, S. Whitcher, & J. Hermanson. 1990. "Guided Imagery and Immune System Function in Normal Subjects: A Summary of Research Findings." In *Mental Imagery,* edited by R. G. Kunzendorf, 179–91. New York: Plenum.

Schröter-Kunhardt, M. 1993. "A Review of Near-Death Experiences." *Journal of Scientific Exploration* 7, no. 3: 219–39.

Scott, R. D. 1978. *Transcendental Misconceptions.* San Diego, Calif.: Beta.

Sebald, H. 1984. "New-age Romanticism: The Quest for an Alternative Lifestyle as a Force of Social Change." *Humboldt Journal of Social Relations* 11, no. 2: 106–27.

Shaffer, J. B. P. 1978. *Humanistic Psychology.* Englewood Cliffs, N.J.: Prentice-Hall.

Shankara. 1947. *Shankara's Crest-Jewel of Discrimination (Vivekachudamani).* Translation and introduction by S. Prabhavananda & C. Isherwood. Hollywood, Calif.: Vedanta.

Shapiro, D. H., Jr., & R. N. Walsh, eds. 1984. *Meditation: Classic and Contemporary Perspectives.* Hawthorne, N.Y.: Aldine.

Sheldrake, R. 1981. *A New Science of Life: The Hypothesis of Formative Causation.* Los Angeles, Calif.: J. P. Tarcher.

———. 1988. *The Presence of the Past: Morphic Resonance and the Habits of Nature.* New York: Random House, Times Books.

Shepard, L., ed. 1991. *Encyclopedia of Occultism and Parapsychology.* 3d ed. Detroit, Mich.: Gale Research.

Silverman, P. S. 1983. "Attributing Mind to Animals: The Role of Intuition." *Journal of Social and Biological Structures* 6:231–47.

Singer, J. L., & J. Kolligian, Jr. 1987. "Personality: Developments in the Study of Private Experience." *Annual Review of Psychology* 38:533–74.

Sloan, D. 1992. "Imagination, Education, and Our Postmodern Possibilities." *ReVision: A Journal of Consciousness and Transformation* 15, no. 2: 42–53.

Smith, J. C. 1976. "Psychotherapeutic Effects of Transcendental Meditation with Controls for Expectation of Relief and Daily Sitting." *Journal of Consulting and Clinical Psychology* 44, no. 4: 630–37.

———. 1987. "Meditation as Psychotherapy: A New Look at the Evidence." In *The Psychology of Meditation,* edited by M. A. West, 136–49. Oxford: Clarendon Press.

Smith, M. B. 1994. "Humanistic Psychology." In *Encyclopedia of Psychology;* edited by R. J. Corsini, 2:176–80. 2d ed. New York: John Wiley & Sons.

Smolensky, P. 1988. "On the Proper Treatment of Connectionism." *Behavioral and Brain Sciences* 11, no. 1: 1–23.

Spangler, D. 1975. *The Laws of Manifestation.* The Park, Forres: Findhorn Foundation.

———. 1977a. *Reflections on the Christ.* The Park, Forres: Findhorn Foundation.

———. 1977b. *Relationship and Identity.* The Park, Forres: Findhorn.

———. 1980. *Explorations: Emerging Aspects of the New Culture.* The Park, Forres: Findhorn.

———. 1984. *Emergence: The Rebirth of the Sacred.* New York: Dell, Delta/Merloyd Lawrence Books.

Spanos, N. P., & P. Moretti. 1988. "Correlates of Mystical and Diabolical Experiences in a Sample of Female University Students." *Journal for the Scientific Study of Religion* 27, no. 1: 105–16.

Spilka, B., R. W. Hood, Jr., & R. L. Gorsuch. 1985. *The Psychology of Religion: An Empirical Approach.* Englewood Cliffs, N.J.: Prentice-Hall.

Stapp, H. P. 1985. "Consciousness and Values in the Quantum Universe." *Foundations of Physics* 15, no. 1: 35–47.

Stearn, J. 1989. *Intimates through Time: Edgar Cayce's Mysteries of Reincarnation.* New York: Penguin, Signet.

Stephenson, J. 1983. *Prophecy on Trial.* Greenwich, Conn.: Trans-Himalaya.

Stevens, B. 1970. *Don't Push the River (It Flows by Itself).* Berkeley, Calif.: Celestial Arts.

———. 1984. *Burst Out Laughing.* Berkeley, Calif.: Celestial Arts.

Stevenson, I. 1993. "Birthmarks and Birth Defects Corresponding to Wounds on Deceased Persons." *Journal of Scientific Exploration* 7, no. 4: 403–16.

Stevenson, I., S. Pasricha, & G. Samararatne. 1988. "Deception and Self-deception in Cases of the Reincarnation Type: Seven Illustrative Cases in Asia." *Journal of the American Society for Psychical Research* 82, no. 1: 1–31.

Sudbery, A. 1986. *Quantum Mechanics and the Particles of Nature: An Outline for Mathematicians.* Cambridge: Cambridge University Press.

Sutherland, S. 1989. *The International Dictionary of Psychology.* New York: Continuum.

Swets, J. A., & R. A. Bjork. 1990. "Enhancing Human Performance: An Evaluation of 'New Age' Techniques Considered by the U.S. Army." *Psychological Science* 1, no. 2: 85–96.

Sykes, J. B., ed. 1976. *The Concise Oxford Dictionary of Current English.* 6th ed. Oxford: Clarendon Press.

Targ, R., & H. E. Puthoff. 1977. *Mind-Reach: Scientists Look at Psychic Ability.* New York: Delacorte Press/Eleanor Friede.

Tart, C. T. 1992. "Perspectives on Scientism, Religion, and Philosophy Provided by Parapsychology." *Journal of Humanistic Psychology* 32, no. 2: 70–100.

Teilhard de Chardin, P. 1959. *The Phenomenon of Man.* New York: Harper & Brothers.

Thomas, G. B., Jr. 1968. *Calculus and Analytic Geometry.* 4th ed. Reading, Mass.: Addison-Wesley.

Thomas, L. E., & P. E. Cooper. 1980. "Incidence and Psychological Correlates of Intense Spiritual Experiences." *Journal of Transpersonal Psychology* 12, no. 1: 75–85.

Thorngate, W. 1990. "The Economy of Attention and the Development of Psychology." *Canadian Psychology/Psychologie canadienne* 31, no. 3: 262–71.

Tiemann, H. A., Jr. 1993. "Archival Data." In *Survey of Social Science: Psychology Series,* edited by F. N. Magill, 293–98. Pasadena, Calif.: Salem.

Turner, E. 1992. "The Reality of Spirits." *ReVision: A Journal of Consciousness and Transformation* 15, no. 1: 28–32.

Tzu, C. 1968. *The Complete Works of Chuang Tzu.* Translated by B. Watson. New York: Columbia University Press.

Ullman, D. 1988. *Homeopathy: Medicine for the Twenty-first Century.* Berkeley, Calif.: North Atlantic.

Ullman, M. 1986. "Access to Dreams." In *Handbook of States of Consciousness,* edited by B. B. Wolman & M. Ullman, 524–52. New York: Van Nostrand Reinhold.

Urantia Foundation. 1955. *The Urantia Book.* Chicago, Ill.: Author.

Vaughan, F. E. 1979a. *Awakening Intuition.* Garden City, N.Y.: Anchor Press/Doubleday, Anchor Books.

———. 1979b. "Transpersonal Psychotherapy: Context, Content and Process." *Journal of Transpersonal Psychology* 11, no. 2: 101–10.

Vitz, P. C., & D. Modesti. 1993. "Social and Psychological Origins of New Age Spirituality." *Journal of Psychology and Christianity* 12, no. 1: 47–57.

Walker, E. H. 1970. "The Nature of Consciousness." *Mathematical Biosciences* 7:131–78.

———. 1977. "Quantum Mechanical Tunneling in Synaptic and Ephaptic Transmission." *International Journal of Quantum Chemistry* 11:103–27.

Wallas, G. 1926. *The Art of Thought.* New York: Harcourt, Brace.

Walsh, R. 1976. "Reflections on Psychotherapy." *Journal of Transpersonal Psychology* 8, no. 2: 100–111.

———. 1984. "Journey beyond Belief." *Journal of Humanistic Psychology* 24, no. 2: 30–65.

———. 1990. *The Spirit of Shamanism.* Los Angeles, Calif.: Jeremy P. Tarcher.

Walsh, R. N., D. Elgin, F. Vaughan, & K. Wilber. 1980. "Paradigms in Collision." In *Beyond Ego: Transpersonal Dimensions in Psychology*, edited by R. N. Walsh & F. Vaughan, 36–53. Los Angeles, Calif.: Jeremy P. Tarcher.

Walsh, R. N., & F. Vaughan. 1980. "A Comparison of Psychotherapies." In *Beyond Ego: Transpersonal Dimensions in Psychology*, edited by R. N. Walsh & F. Vaughan, 165–75. Los Angeles, Calif.: Jeremy P. Tarcher.

———. 1993a. "The Art of Transcendence: An Introduction to Common Elements of Transpersonal Practices." *Journal of Transpersonal Psychology* 25, no. 1: 1–9.

Walsh, R., & F. Vaughan, eds. 1993b. *Paths beyond Ego: The Transpersonal Vision.* Los Angeles, Calif.: J. P. Tarcher; Perigee.

Watkins, J. G., & H. H. Watkins. 1986. "Hypnosis, Multiple Personality, and Ego States as Altered States of Consciousness." In *Handbook of States of Consciousness*, edited by B. B. Wolman & M. Ullman, 133–58. New York: Van Nostrand Reinhold.

West, M. A. 1987. "Traditional and Psychological Perspectives on Meditation." In *The Psychology of Meditation*, edited by M. A. West, 5–22. Oxford: Clarendon Press.

Westrum, R., ed. 1992. *Social Psychology of Science,* September.

Wheeler, J. A. 1983. "Law without Law." In *Quantum Theory and Measurement,* edited by J. A. Wheeler & W. H. Zurek, 182–213. Princeton, N.J.: Princeton University Press.

Wheeler, J. A., and W. H. Zurek, eds. 1983. *Quantum Theory and Measurement.* Princeton, N.J.: Princeton University Press.

White, P. A. 1982. "Beliefs about Conscious Experience." In *Awareness and Self-awareness.* Vol. 3 of *Aspects of Consciousness,* edited by G. Underwood, 1–25. New York: Academic.

———. 1988. "Knowing More about What We Can Tell: 'Introspective Access' and Causal Report Accuracy Ten Years Later." *British Journal of Psychology* 79:13–45.

Wigner, E. P. 1983. "Remarks on the Mind-Body Question." In *Quantum Theory and Measurement,* edited by J. A. Wheeler & W. H. Zurek, 168–81. Princeton, N.J.: Princeton University Press.

Wilber, K. 1975. "Psychologia Perennis: The Spectrum of Consciousness." *Journal of Transpersonal Psychology* 7, no. 2: 105–32.

———. 1979. "A Developmental View of Consciousness." *Journal of Transpersonal Psychology* 11, no. 1: 1–21.

———. 1980. *The Atman Project: A Transpersonal View of Human Development.* Wheaton, Ill.: Theosophical Publishing House.

———. 1982. "Physics, Mysticism, and the New Holographic Paradigm: A Critical Appraisal." In *The Holographic Paradigm and Other Paradoxes: Exploring the Leading Edge of Science,* edited by K. Wilber, 157–86. Boulder, Colo.: Shambhala.

Wilhelm, R., & C. F. Baynes, trans. 1967. *The I Ching or Book of Changes.* 1950. Reprint, Princeton, N.J.: Princeton University Press.

Wilson, R. A. 1988. "The Priory of Sion." *Gnosis: A Journal of the Western Inner Traditions* 6:30–39.

Wise, R. A., & P.-P. Rompre. 1989. "Brain Dopamine and Reward." *Annual Review of Psychology* 40:191–225.

Wolff, F. F. 1939. "Concept, Percept, and Reality." *Philosophical Review* 48:398–414.

Wren-Lewis, J. 1988. "The Darkness of God: A Personal Report on Consciousness Transformation through an Encounter with Death." *Journal of Humanistic Psychology* 28, no. 2: 105–22.

Yogananda, P. 1981. *Autobiography of a Yogi.* 12th ed. Los Angeles, Calif.: Self-Realization Fellowship.

Young, A. M. 1976a. *The Geometry of Meaning.* Mill Valley, Calif.: Robert Briggs.

———. 1976b. *The Reflexive Universe: Evolution of Consciousness.* Mill Valley, Calif.: Robert Briggs.

Younger, J., W. Adriance, & R. J. Berger. 1975. "Sleep during Transcendental Meditation." *Perceptual and Motor Skills* 40:953–54.

Yram. 1967. *Practical Astral Projection.* New York: Samuel Weiser.

Zimmerman, D. W. 1984. "A Note on the Completeness of the Scientific Method." *Psychological Record* 34:175–79.

## INDEX

**F**



**G**



**H**



**I**